Pro Tools Bible

David Leathers

McGraw-Hill

New York Chicago San Francisco Lisbon
London Madrid Mexico City Milan New Delhi
San Juan Seoul Singapore Sydney Toronto

The McGraw·Hill Companies

**Cataloging-in-Publication Data is on file with the
Library of Congress.**

Copyright © 2004 by The McGraw-Hill Companies, Inc. All rights reserved.
Printed in the United States of America. Except as permitted under the
United States Copyright Act of 1976, no part of this publication may be
reproduced or distributed in any form or by any means, or stored in a data
base or retrieval system, without the prior written permission of the pub-
lisher.

1 2 3 4 5 6 7 8 9 0   DOC/DOC   0 9 8 7 6 5 4 3

ISBN 0-07-141234-4

*The sponsoring editor for this book was Steve Chapman and the production
supervisor was Pamela Pelton. It was set in Century Schoolbook by MacAllister
Publishing Services, LLC.*

Printed and bound by RR Donnelley.

 This book is printed on recycled, acid-free paper containing a minimum of 50
percent recycled de-inked fiber.

McGraw-Hill books are available at special quantity discounts to use as premiums
and sales promotions, or for use in corporate training programs. For more information, please write to the Director of Special Sales, Professional Publishing, McGraw-
Hill, Two Penn Plaza, New York, NY 10121-2298. Or contact your local bookstore.

To Debbie, Phil, and Emily

CONTENTS

Contents

Contents

Contents

PREFACE

This book is a resource for Pro Tools users. It has three main sections: a section of tutorials, a hardware section, and a section that lists and describes all the incredible plug-ins and compatible software from Digidesign and other companies that form a hugely powerful creative environment. The tutorials are designed to orient both new and experienced users to Pro Tools 6.x.

In addition to the tutorials and a lot of information about Pro Tools systems, this book is also a research guide that will dramatically shorten the time you spend sorting out the information overload about plug-ins and other software. I've done the legwork for you. If you have attempted much of this research on your own, you know how tedious it can be. I'm a pretty good *audiovisual* (AV) journalist with a fair amount of connections, and it still took me a year to pull this together. So, if you are hungry for this kind of information, this book is for you. It contains URLs for over 200 Pro Tools-compatible products, most of which have free, downloadable software compatible with Pro Tools systems. These links are arranged by their functional categories rather than by the companies that manufacture them. This arrangement should make it easy for you to get to the information and products you are interested in without having to sort through endless web sites and magazines. For instance, if you need to find a reverb plug-in for your system, go to the reverb section of this book. You will find every Pro Tools-compatible reverb software in one place. The descriptions will tell you about each product and give you a web address where you can get the most up-to-date information, and often a free demo or limited-time version.

We strongly considered including a disk with some of the demos, but rather than delay publishing or make it more expensive, we decided it made more sense to just provide end users with the information and URLs. This way you can get what you want when you want it, and it will be current.

I've been working with Pro Tools for years in both music recording and audio post. I've talked to lots of manufacturers and Pro Tools users, been to a lot of trade shows and demonstrations, searched every web site, and spent a lot of time directly analyzing products. It's a competitive world for audio software and you are their customer base. The best and smartest of the manufacturers tend to want to let you try their stuff out and have a look before you buy. As you will see, there is a universe of development out there and it is just a few clicks away. This book is an excellent guide to that universe and is designed to make exploring it a pleasant and productive experience.

ACKNOWLEDGMENTS

Special thanks to Brian McKernan and Tim Tully for help in getting started and guidance in the project; to Steve Chapman from McGraw-Hill and Andrea Koehler from MacAllister Publishing Services for all their help in managing and editing the book; to Ed Gray, Joe Saracino, Chandra Lynn, and many others at Digidesign for all their help and support; and to Benjamin Chadabe at GRM Tools for valuable help, insights, and encouragement.

Also, thanks to the many musicians, filmmakers, producers, and engineers that I have had the pleasure of working with and learning from over the years.

Learning and Using Pro Tools

CHAPTER 1

Getting Started

Pro Tools is a deep, complex program. Most likely, you have specific things in mind that you are interested in doing with the system and will approach it from a perspective that is oriented toward those tasks. This goal could be anything from recording sound ideas in a personal project studio to running a large-scale audio post system doing mixes for movie soundtracks. Different types of projects require different Pro Tools skills and different utilization of specific Pro Tools features. But, regardless of your intended use, some basic concepts and procedures can be adapted and used in a wide variety of ways.

The exercises in this book are designed for beginning Pro Tools users as well as for users who have some experience and want to explore the system further. However, it is not a tutorial on everything about audio and computers. The assumption is that you have some familiarity with using your computer and some experience with recording and mixing audio. These exercises will quickly immerse you in various Pro Tools procedures and expose you to fundamental concepts. There will be plenty to think about as you work through them. Take your time.

Mistakes and dead ends are inevitable in the beginning stages of the learning curve. Using a program as a beginner can also be frustrating because it is hard to be as productive as you want to be while you are still trying to remember the location of controls and the sequences of procedures for many different operations.

For instance, as an experienced video editor, you may be very good at using the audio tracks in Avid systems, Media100, or Final Cut Pro. The simple editing of audio tracks may be easy for you if you just continue using a program you have used hundreds of times. However, if you can take the time and get used to doing the same work in Pro Tools, you will be on the way to understanding the program and will be able to access the advanced Pro Tools capabilities for your project. All of these programs do many of the same basic things with audio tracks, but Pro Tools gives you a very deep audio feature set that ultimately lets you do much more than you can in even the most full-featured video-editing program.

The situation may be similar if you are a musician or engineer who is used to working with analog recording equipment. Having to relate to the computer screen and to nonlinear access to recording and editing may seem foreign at first. But once you are over that hump, the benefits will start to outweigh the problems.

If you are already familiar with another nonlinear digital audio workstation, such as Emagic Logic or MOTU Digital Performer, then you already understand most of the basic concepts of Pro Tools. Adapting to Pro Tools will simply be a matter of learning the specific characteristics of this system as opposed to the others.

Before You Start

Before you start, take a look through the extensive documentation that has been provided by Digidesign. You may have hard copies of some of them, at least the Getting Started manual and the Pro Tools reference guide with Pro Tools *time division multiplexing* (TDM) software. With the *limited edition* (LE) software, your manuals are going to be PDF files on the install disk. Even with TDM software, there are a number of additional manuals in PDF form on the disk for which you won't have hard copies.

If you have any way at all to do it, it's very worthwhile to print out all of the documentation and put it in a binder. This is a little expensive and certainly a hassle, but its more likely that you'll read the material this way. Its easier to refer to the hard copy manuals when you're using the system than to open Acrobat Reader or a help program simultaneously. The screen can get cluttered enough in normal operation without the onscreen manuals on top of the interface you are working with. You really want to keep the screen real estate available for what you're actually doing with Pro Tools, if you can. If you must use the PDF files in an onscreen format, the next best alternative is to use a second computer to display the manuals. For example, if you are working on a desktop system and have a laptop, you might use the laptop for the manuals.

Even if you never read manuals and you simply want to get started, it's still a good idea to get hard copies of the manuals and check them out. It may not be necessary to read all the way through them, but at least browse them and get the feel for the information. It will shed some light on the nature and depth of what you're undertaking. If you are a fairly new user, your understanding of the program, what it is and what it does, will change over time as your familiarity grows. Referring back to the manuals will help you explore and understand the details as you have new experiences with the program's features and characteristics.

The Getting Started manual is pretty much a "must read." It's going to tell you how to make sure your system is set up correctly at the most basic level. You really do need to go through this process and understand the way the hardware is set up. This knowledge will continue to be a factor in your ability to understand the system as a whole and to get it to work the way you want it to.

Once you get into the software, you have to remember that Pro Tools is designed to accomplish many tasks, many of which are very similar, with only subtle differences. Pro Tools has seen a lot of development since its introduction more than a decade ago and, as the designers have become more knowledgeable, the program has become very flexible, with multiple ways to accomplish the same things. If you are a musician, or if you've ever tried to learn a musical instrument, think of it like the concept of alternate fingerings, where there may be more than one way to play a note or a chord and the one you choose is influenced by what you are trying to do, what you were doing just before that, or what you plan to do next. Pro Tools can be much the same. There may be several factors in your choice of procedures. The bottom line is not always doing it right as much as finding a way to do it that fits your workflow and provides the results you want.

As well developed as Pro Tools is, there are still bound to be some system-based problems and challenges, and there are definitely going to be operator errors and mistakes. Part of learning Pro Tools is learning how to avoid problems and to create solutions or work-arounds when needed. That takes time and practice.

A lot of safeguards are built into Pro Tools, based on very practical experience. This program has been used by all kinds of music and postproduction professionals working very long hours, very late at night, under extreme pressure, and, no doubt, in various states of mind. These serious users have been providing feedback to Digidesign for years, and the developers have used this information to constantly improve the technology. Because this develop-use-feedback-develop-again cycle has been going on for so long, the program has become fairly idiot-proof. But there are a lot of subtleties to understand if you want to get the most out of the program.

Now that I've given you all the disclaimers, it's also important to remember that there is a big payoff for spending the time and energy to learn as much as you can. It's like learning how to use any high-powered machine. The more control you get, the more you will realize what the system can really do and the less afraid you are to use it.

The Exercises

The exercises in this book will walk you through key parts of the interface and some basic workflow procedures. They are intended to immerse you in common processes and illustrate a few key concepts in order to provide a foundation from which you can go forward. Ultimately, becoming fluid in applying the system's logic to real-world tasks requires a lot of repetition.

The exercises assume you have a basic understanding of the computer and of audio recording. You should be familiar with basic microphone techniques, setting sound and signal levels, and how multitrack recording and mixing are used to create audio. There are many excellent books on specific aspects of recording, engineering, and mixing music that you can access to explore the vast body of knowledge on these subjects. What follows is very Pro Tools specific and very basic.

You will be working with your own audio sources. If you are not a musician, then you will need to find one to play a few tracks for you. All you need is some form of percussion instrument and a guitar or keyboard instrument. Some vocals would be nice, too. If you don't want to work with vocals, substitute a lead guitar or some other melody instrument. However, you can use any other instruments that you may have available. It's best to do the exercises in order, since each one builds on the previous exercise's work.

Macintosh Versus PC

With Version 6.1 of the software, the first OS X-Windows XP cross-platform compatible version, Pro Tools is basically identical in the Macintosh OS X and Windows XP environments. The exercises in this book are done on a Macintosh, and where there are some differences for a PC, we'll point those out. There are the usual differences in some basic functions on the computer keyboard. Specifically, the Macintosh command key equals the PC control key. The Macintosh control key equals the PC start key. The Macintosh option key equals the PC alt key. The Macintosh return key equals the PC enter key.

Exercise 1-1: Setting Up a Session

First, at the most basic level, let's go through a few steps and look at what it takes to set up a Pro Tools session.

You should have Pro Tools installed and your studio set up. You should have your audio sources connected to Pro Tools inputs. You should have a monitoring system connected to your Pro Tools outputs. You should also be able to monitor your system's output on headphones with the speaker volume turned down or off. If you are going to be using *musical instrument digital interface* (MIDI), you should have your MIDI interface and subsystems set up and connected to your computer.

Before launching Pro Tools, make sure that your audio interfaces, MIDI interfaces, and MIDI devices are turned on. If this is not the first time you have opened Pro Tools 6, then you should trash the preferences. The Preferences folder is located in Users/Your_User_Name/ Library/Preferences in Macintosh OS X. Take the Pro Tools Preferences file and drag it to the trash. Trashing the preferences (with the application closed) and then restarting the program will cause Pro Tools to resume the default settings.

When you installed Pro Tools in OS X, it should have been installed in a new Digidesign folder in the Applications folder on your main hard drive. It's a good idea to put an alias for the Pro Tools application in the dock. In OS X, you don't need to make the alias first, as you did in OS 9. In OS X, you do this simply by dragging the application icon to the dock. An alias will be automatically created and placed in the dock. The actual Pro Tools application will stay where it is.

Now, click on the new icon in the dock and launch Pro Tools. At first, all you will see will be the Pro Tools menus across the top of an empty desktop. The first thing you have to do is either open an existing Pro Tools session or create a new one.

New Session Dialog

In order to start a new session, you have to go the New Session dialog box. Use the following steps:

1. Go to the pull-down menu, File > New Session, or use the keyboard equivalent, which is command (control) + N (see Figure 1-1).

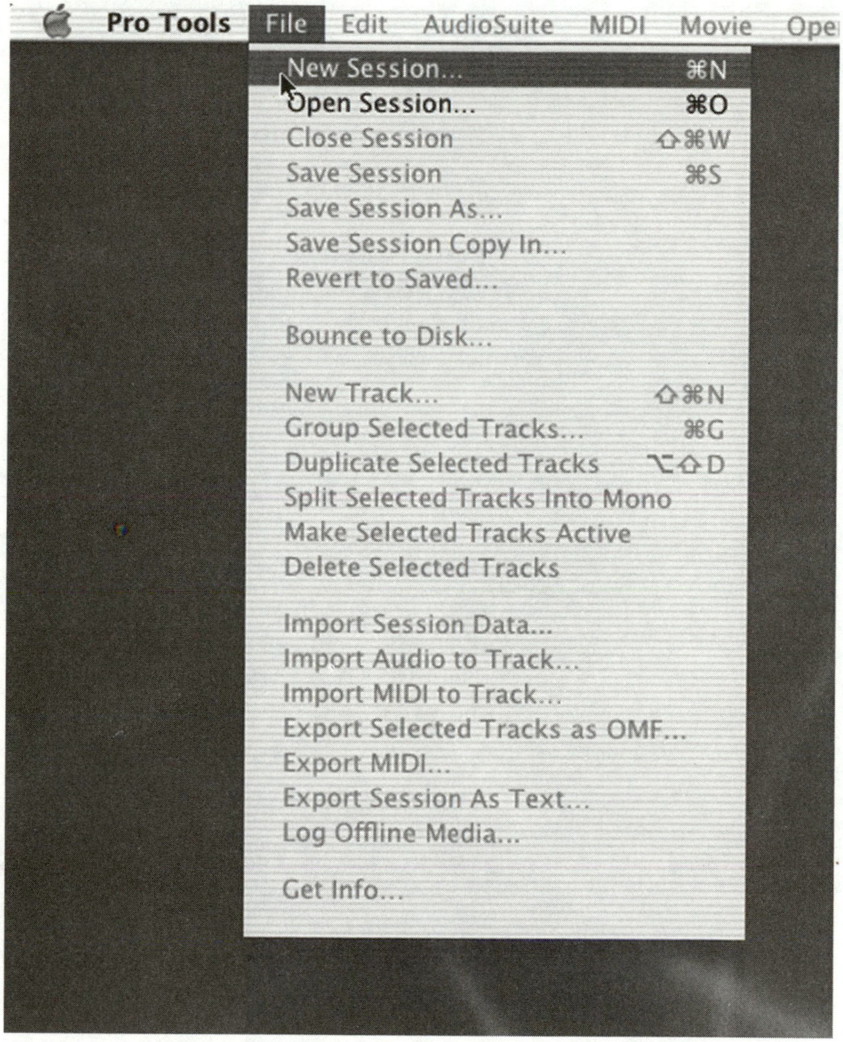

Figure 1-1
Starting a new
Pro Tools Session

Figure 1-1
Starting a new
Pro Tools Session

This action will open the New Session dialog box (see Figure 1-2).

A dialog box is a type of software interface component that generally requires you to provide information. It asks questions and you answer, hence the name *dialog*.

Figure 1-2
The New
Session dialog

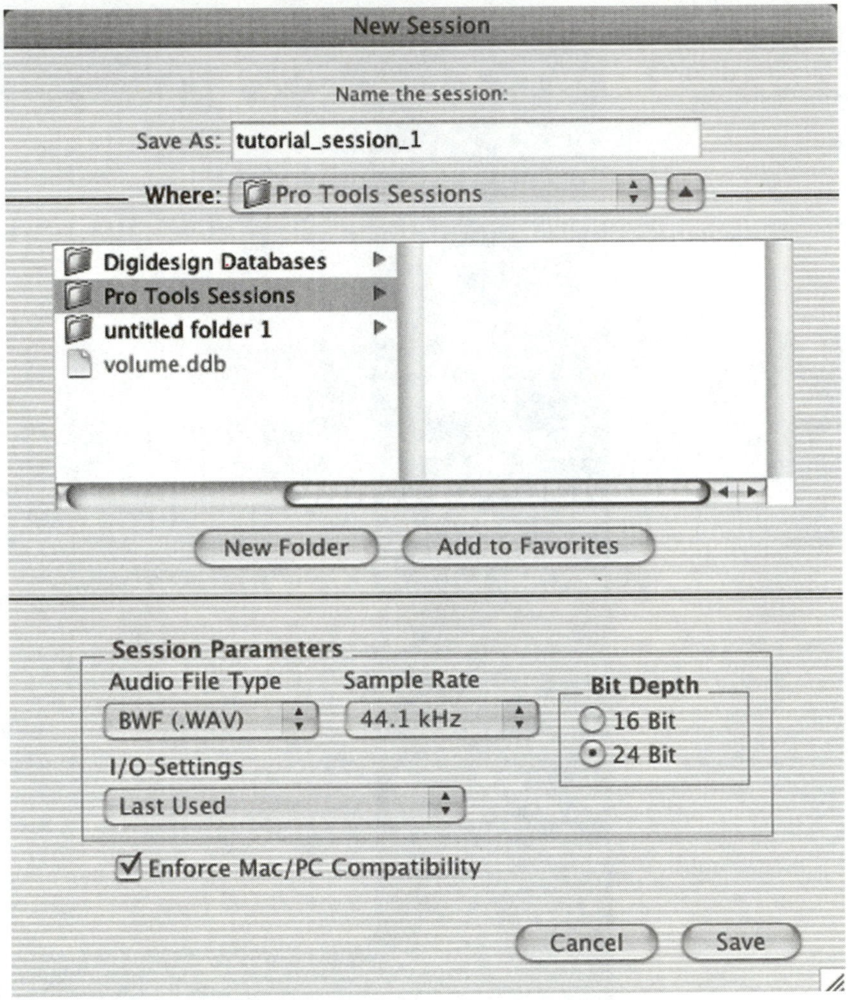

In this dialog box, you will be making several important decisions that will affect the rest of your project. The first is the name of the project itself.

2. In the New Session dialog box, type the name of your session in the Save As field. Name this session "tutorial_session_1." (Different computer file systems

use spaces, dashes, slashes, and other symbols in different ways. Using underscores instead of spaces between words in the manner shown is a UNIX convention and is a fairly safe way to ensure that names will stay intact when transported between unlike systems. Since OS X is really a UNIX-based system, it makes sense to follow this convention.)

Next, you have to indicate where you want the project stored. The project should be stored on a drive other than your main system drive. Several kinds of drives are suitable for this purpose, including some kinds of internal drives, external FireWire drives, *small computer systems interface* (SCSI) drives, *redundant array of independent disks* (RAIDs), or *storage area networks* (SANs), as described in the hardware chapter of this book. For most project studios, having at least one external FireWire drive is a very convenient addition to your system. It makes the whole project portable in a very small package and at this point they are inexpensive and quiet enough to have in the room. They also generally have a very small footprint. So, in the dialog box, navigate to your audio drive.

3. Using the New Folder button, create a new folder. Name the new folder "Pro Tools Sessions." Set the session parameters as follows.

Session Parameters   In the area marked Session Parameters, you are offered several audio file types, sample rates, and bit depths. For this session, we will stick with the default bwf (.wav) file type.

About Sample Rates   For sample rates, all Pro Tools systems have the option of the 44.1 kHz sampling rate. Pro

Tools LE and TDM systems will include the option of a 48 kHz sampling rate. Pro Tools LE with DIGI 002 will include the 96 kHz option. Pro Tools HD systems, depending on the configuration, may also include a 192 kHz sampling rate option. For the bit depth, you have a choice of 16 bits or 24 bits on appropriately configured systems.

If you are creating a music project that will likely end up on an audio CD, the 44.1 kHz sample rate is probably the right choice, as that is the sample rate for standard audio CDs. In audio postproduction for video and film, 48 kHz is generally the standard used. It's the standard for AES audio on Digital Betacam and also is the recording sample rate used by many *digital video* (DV) camcorders. By choosing the right sampling rate, you avoid sample rate conversions later on, which tend to have a detrimental effect on the sound.

About Bit Depth   For the Bit Depth column, you can choose 16 bits or 24 bits. A depth of 24 bits requires 50 percent more disk space than a depth of 16 bits, but there is a significant advantage sound wise. When you are at the end of a 24-bit music project and want to create an audio CD, part of the mastering process will involve dithering, which is the process of converting your 24-bit file to the CD standard of 16 bits at 44.1 kHz. You will generally be better off working at 24 bits and dithering down to 16 bits for mastering than if you work at 16 bits all the way through. Many engineers feel that this actually improves the sound significantly. So, if your system has a 24-bit interface, select 24 bits. If not, select 16 bits.

For now, leave the *input/output* (I/O) settings at "Stereo Mix" and leave the Enforce Mac/PC Compatibility box checked, which is the default.

4. Click Save at the bottom of the dialog box.

Once you click the Save button, two things will happen. First, Pro Tools will automatically create a new session folder with the same name as your session. It will place your session file inside the folder. Two new folders, the Audio Files folder and the Fade Files folder, will also be automatically created and placed inside the session folder (see Figure 1-3).

Second, the Pro Tools session will open, revealing the three main Pro Tools windows: the Edit window, the Mix window, and the Transport window. The Edit window and the Mix window are interactive and show much of the same information in different ways and have different controls.

Figure 1-3
The session folder

The Setup Menus

Once you have created your session, you may want to jump right in and start recording. But you are actually better off if you do a little more housekeeping first.

1. Go to the pull-down menu, Setups > Hardware Setup. The Hardware Setup dialog box opens. The contents of the Hardware Setup dialog box will vary depending on the hardware in your system. There are fields to control the parameters of your hardware interfaces, most of which are self-explanatory.

2. Check and make sure that your interfaces are set up and recognized by the system.

3. Close the Hardware Setup dialog box.

4. Go to the pull-down menu, Setups > I/O Setup. The I/O Setup dialog box will open and give you a graphical picture of the inputs and outputs of your system.

5. Click on the Input tab and you will see how your inputs are presently configured (see Figure 1-4).

The contents of this window will appear differently depending on your system's I/O hardware. If you have an Mbox, you will see two analog inputs and a *Sony/Phillips digital interface* (S/PDIF). If you have a DIGI 001, you will see eight analog inputs, eight ADAT inputs, and two channels of S/PDIF. For the purposes of this tutorial, we will use a single eight-channel analog interface (888/24 I/O) for input and output. Similarly, if you click on the Output tab, you will see the way your outputs are configured. The labels are shown in stereo pairs. You can open them up to reveal the mono inputs and outputs by clicking on the arrow to the left of the stereo pair. Open them all up now (see Figure 1-5).

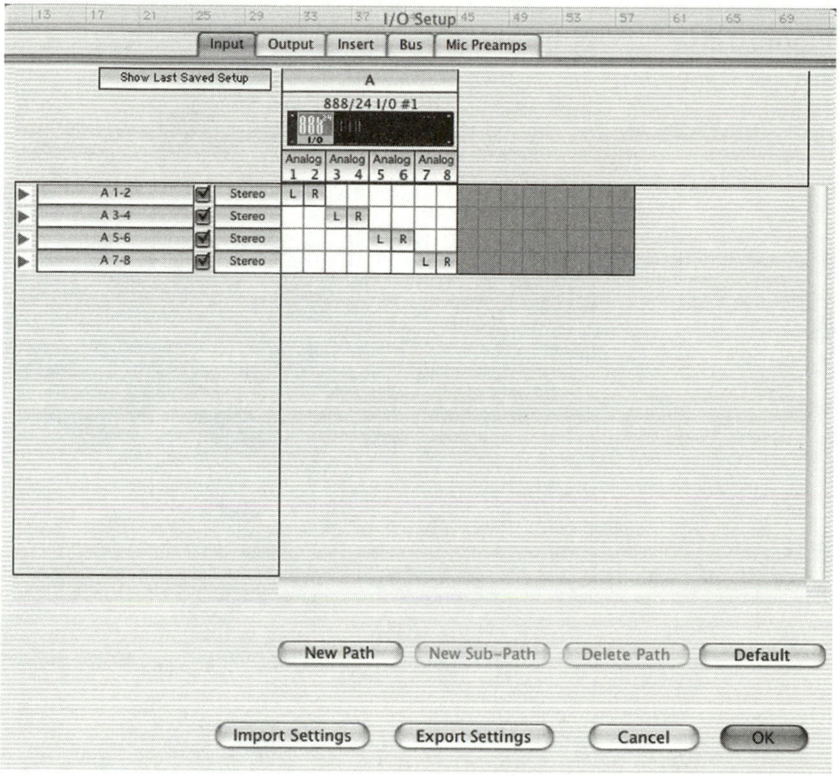

Figure 1-4

Stereo pairs in
the I/O window

Figure 1-4
Stereo pairs in
the I/O window

The default labeling for inputs and outputs varies depending on the hardware configuration and can be confusing. It's a good idea to rename both the inputs and the outputs to reflect what they actually are. In other words, take all the inputs and label them "IN 1–2, IN 1, IN 2, IN 3–4, IN 3, IN 4, IN 5–6," and "OUT 1–2, OUT 1, OUT 2," and so on.

6. With the Input tab selected, double-click on each input name and type in the new name.

7. Select the Output tab and do the same for the outputs. Once you have renamed your inputs and outputs, close the I/O Setup dialog box by clicking OK.

Figure 1-5
The open stereo
pairs

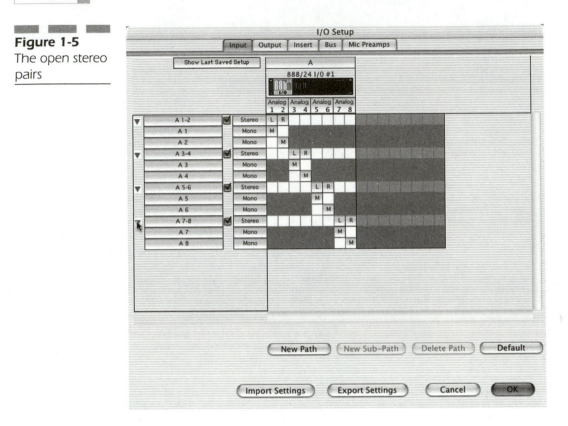

The Edit Window

The Edit window is in many ways the most important window in Pro Tools. It is where you will spend most of your time and where you generally have the best overview of your project. When you first open the project, you will have an empty Edit window.

The first thing to do is expand the Edit window by dragging the lower right-hand corner of the window so that it goes almost all the way across the screen and almost to the bottom of the screen. If you have two screens, it's a good idea to move the Mix window to the second screen. Otherwise, it can be hidden under the Edit window and brought up at any time by using the Window > Show Mix pull-down menu. You can put the Transport window near the bottom where it will be out of the way. You will notice that the Transport window always floats above the other windows so it's not going to get lost, but it does get in the way. As you get into Pro Tools, you will find that you will want to use keyboard equivalents to do most of the operations that are represented in the Transport window functions anyway.

In order to simplify the interface for now, go to the pull-down menu Display > Ruler View Shows and select "None."

Then repeat the Display > Ruler View Shows pull-down and select Bars:Beats. This will eliminate all the rulers at the top of the Edit window except the Min:Sec ruler and the Bars:Beats ruler. The others can be called up at any time by using the pull-down menu, but you don't need them all the time and hiding them makes it easier to read the other rulers that you are going to use.

New Track Dialog

Even though you are in the Edit window, there still aren't any audio tracks to record on. For this exercise, let's start with three audio tracks.

1. Go to the File > New Track pull-down menu to open the New Track dialog box.

2. Use the default values and click the Create button or hit return (enter on a PC). You will now see one audio track in the Edit window. Use the Window > Show Mix pull-down menu to look at the Mix window and you

will also see the new track there. Now click anywhere on the Edit window and it will again come to the front.

3. To create the other two tracks, use the keyboard equivalent command (control)-shift + N to reopen the New Track dialog box.

4. This time type **2** into the Create field and click the Create button. Two more tracks will appear in both the Edit and Mix windows.

Saving the Session

It is important to save your sessions often. Use the File > Save Sessions pull-down menu or the keyboard equivalent command (control) + S to save the session.

Even though crashes are much less frequent with OS X than with previous Macintosh operating systems, it is possible to freeze the program occasionally. If this happens, you will have to force-quit the application, and all the unsaved changes will be lost. It is a good idea to enable the AutoSave function by using the Setups > Preferences pull-down menu to access the Preferences dialog box and clicking the Enable Session File Auto Backup button under the Operations tab. This will keep you from losing more than a few minutes' worth of work.

CHAPTER 2

Recording Tracks

Naming Tracks

For this exercise, you will record a drum machine (or other rhythm source) to the first track; a guitar, keyboard, or other harmonic instrument to the second track; and a vocal, horn, lead guitar, or other melodic instrument to the third track.

1. In the Edit window, double-click on the field Audio 1 in the first track. This opens up the dialog box for the track and the name Audio 1 is highlighted. Type drums, or another term descriptive of your audio source, into the Name of Track box.

2. Click OK to close the box.

3. In the same manner, rename the other two tracks Guitar 1 and Vocal 1, or other descriptive names.

4. Use the keyboard equivalent command (control) + S to save the session with the name changes.

You will notice that the new names appear in the Show/Hide area just to the left of the tracks in the Edit window. Take a look at the Mix window and you will see that the new track names appear there as well.

In the Edit window, in the Show/Hide area on the left, click on the names of the second two tracks. This click will hide the tracks from view while we work on recording the drums. When you want to see those tracks again later, just go back to the Show/Hide area and click on the tracks you want to see.

Exercise 2-1: Recording a Track

For the first track of this session, let's start with the drum track. For your audio source, use a drum machine or synthesizer that can output a drum pattern. If neither is available, use a metronome with a microphone or you can beat out a pattern on any percussion instrument and use a microphone. Keep the pattern simple. This is not going to be a masterpiece. It's just going to illustrate a few features of the system. To record the drum track, use the following steps:

1. Use the pull-down menu, Display > Edit Window Shows > I/O View, to show the I/O view in the Edit window. You should see that the input to Track 1 is set to IN 1.

2. Now make sure that your studio monitors are turned down very low until you are sure that sound is coming through the system properly. Make sure the output of your sound source is connected to the input of Channel 1.

3. Put the track into record mode by clicking the R. The track is now armed to record. If you play your audio source, you should see it on the audio meters in the track and hear it through your monitors. Make sure you are getting a strong signal that is just getting into the yellow part of the meter but is not clipping.

4. Click the Record button (the red one) in the Transport window and then click the Play button in the Transport window to begin recording. Record a minute

or two of your drum track and then hit the spacebar to stop recording. Congratulations! You just recorded some audio.

5. Take the track out of record mode by clicking the R button in the drum track.

6. Hit the return (enter) key to move the playhead back to the beginning of the session.

7. Hit the spacebar to play back and hear what you recorded. It should look something like Figure 2-1.

Aborting or Deleting Recorded Takes

If you decide that you don't like something you are recording and want to start over, you can press command (control) + period (.) at any time during the recording process. Recording will stop immediately and the audio file will not be saved. If you want to delete the file immediately after recording has stopped, you can use the pull-down menu, Edit > Undo Record Audio, or the keyboard equivalent, command (control) + Z.

Figure 2-1
The drum track

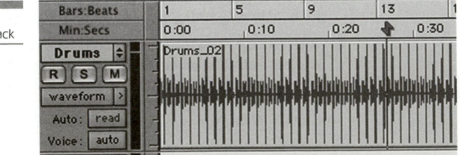

Audio Files and Audio Regions

Understanding the difference between audio files and audio regions is one of the most important concepts in grasping the program.

The audio you just recorded is now stored as an audio file. It resides in the Audio Files folder that was automatically created in the tutorial_session_1 folder when you created the session. Each audio file contains the actual digital audio samples in the .WAV format that we chose in the New Session dialog box when we created the session. What you see in the Pro Tools timeline is not the audio file, but the audio region. The audio region is a separate pointer file that provides a graphic representation of the audio file and is stored as part of the session file. When you edit an audio region in Pro Tools, you are not editing or changing the audio file. It stays completely intact, just as you recorded it. You are editing the audio region only. The audio region is just a part of a script that tells Pro Tools what parts of the audio files to play back and when. When you play back your Pro Tools session, Pro Tools reads the audio regions and plays back only the parts of the audio files that are represented by the edited audio regions in the tracks. So, by editing audio regions, you are actually scripting the way that Pro tools will selectively play back parts of audio files.

In the next section, we will cut out a small section of an audio region using Pro Tools' editing tools. Next we will use a Copy command to copy that audio region. Then we will paste it back into the track to make it repeat itself many times. The cut and paste operations are done the same way as they are done in a word processor. In this case, you will place many identical audio regions in the track, but there is still only one audio file that gets played back over and over when Pro Tools is put into play mode. As the playhead passes over the identical audio regions in succession, Pro Tools plays back the same part of the audio file over and over, precisely conforming to the specifications defined in the edited audio regions.

Saving Versions of a Session

At this point, it makes sense to save the session and to start working on a new session with a new name. We're going to try building a different drum track from a short loop. If you want to get back to exactly this point and use your first drum track later, you can simply close your new session and reopen this one.

1. Use the File > Save Session pull-down menu or the keyboard equivalent, command (control) + S, to save the session as you did before.

2. After you have saved the session, use the File > Save Session As pull-down menu to open the Save Session As dialog box. This time, name the session tutorial_session_1a. Now, from this point forward, you will be working on Session 1a and the previous Session 1 will remain intact on your hard drive. If, after working on the new version for a while, you are unsatisfied with what comes next, you can always go back to the first version and start from there again.

The Edit Tools

Before we go on, it's time to look at the edit tools. These are the main tools that you will use all the time in Pro Tools. They are located above the rulers (see Figure 2-2).

Figure 2-2
The edit tools
buttons

For now, we will just look at the basic functions of the tools that we will be using in the next exercises. Some of them have additional functions. For all the details, refer to your reference guide.

The Zoomer Tool

With the Zoomer tool, you can zoom into or out of any area of a track (see Figure 2-3). You will zoom in when you want to see greater detail and zoom out when you need to see the wider context you are working in. With the Zoomer tool selected, simply move the cursor over the point you want to zoom in on and click. It will zoom in one zoom level and center your target point. Keep clicking to move in further. To zoom out, option-click (alt-click).

The Trimmer Tool

The standard Trimmer tool is used to shorten or lengthen audio regions (see Figure 2-4). It is nondestructive in that

Figure 2-3
The Zoomer tool

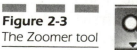

it does not change the actual audio file. With the Trimmer tool selected, move the cursor near the start or end of an audio region. The Trim cursor will appear. Hold down the mouse and drag the Trim cursor in either direction. The audio region will be shortened or lengthened accordingly. Eventually, as you drag the end of the audio region to the right, extension will stop when you reach the end of the source audio file. You can extend a region only if there is an actual audio file to support it.

The Selector Tool

The Selector tool is used to select parts of audio regions for editing or playback (see Figure 2-5). With the Selector tool selected, move the cursor over an audio region. It will turn into an I. Click and drag in either direction to select part of an audio region. You can also drag across multiple tracks with this tool, creating multitrack selections.

The Grabber Tool

The Grabber tool is used to select and move entire audio regions (see Figure 2-6). Move the cursor over an audio region. The cursor turns into a hand. Click once and the audio region is selected. If you click and hold the mouse down, you can drag the audio region forward or backward in a track or to a completely different track.

Figure 2-5
The Selector tool

Figure 2-6
The Grabber
tool

The Scrubber Tool

The Scrubber tool enables audio scrubbing (see Figure 2-7). Audio scrubbing means that you can hear the audio on one or two adjacent tracks by dragging the Scrubber tool back and forth across a section of the track. This can be useful in trying to locate a particular sound in a track. When you stop the Scrubber tool at a particular point in an audio region, the playhead cursor stays at that point.

The Smart Tool

The Smart tool is activated when you click on the oval just below the Selector tool (see Figure 2-8). It lets you use the Trimmer tool, the Selector tool, and the Grabber tool (and also create crossfades) without having to switch back and

Figure 2-7
The Scrubber
tool

Figure 2-8
The Smart tool

forth. With the Smart tool activated, as you move the cursor around an audio region, the tool will change, depending on where the cursor is. In the middle of an audio region, the cursor will be the Selector tool in the top half of the region and the Grabber tool in the bottom half of the region. Near either end of the audio region the cursor will become a Crossfade tool in the top half and the Trimmer tool in the bottom half.

The Pencil Tool

Pro Tools audio regions are graphic representations of sound. With the Pencil tool you can draw in graphic changes that will then be reflected in the sounds represented (see Figure 2-9). The Pencil tool is a destructive tool. This means that it permanently changes the audio file when you make changes to the audio region. There are many uses for the Pencil tool. One example is that you can use it to repair pops or clicks in the audio by zooming in all the way to the sample level of the waveform and simply drawing the pops out.

Figure 2-9
The Pencil tool

Edit Modes

The behavior of the tools and the tracks changes depending on what edit mode you are working in. You choose your edit mode by selecting one of the four buttons in the upper-left corner of the interface (see Figure 2-10). The default edit mode is Slip, which allows the tools to work freely and fluently. Grid, Spot, and Shuffle modes restrict movement in ways that will help position audio regions in a variety of special circumstances. Refer to your reference manual for a complete explanation of edit modes.

Exercise 2-2: Creating a Drum Track by Looping

This time, you will build the drum track by creating a simple drum loop and then using the loop repeatedly in the track. This procedure will introduce several other features of the system. As discussed earlier, in Pro Tools there are many ways to accomplish similar results. Don't worry about the artistic quality of your results so much at this point. We are just going to play with a few more tools so you may consider what you can do with them.

1. Delete the audio in the drum track by highlighting the clip with the Grabber tool and then pressing the

Figure 2-10
The edit mode
buttons

delete key. You have now deleted the audio region from this session. However, it still exists in your previous session if you ever want to go back to it.

2. Press the return (enter) key to move the playhead to the beginning of the track.

3. Now, as before, make sure that your studio monitors are turned down fairly low. Put the track into record mode by clicking the R in the drum track. This time, use the F12 key to begin recording immediately.

4. Record a few bars of your drum pattern. Hit the spacebar to stop recording. You should see some audio in the track represented as a waveform (see Figure 2-11).

5. Click the R again to take the track out of record.

6. Press the Return to Zero button in the Transport window.

7. Now press the spacebar to play the timeline. You should hear your drum part being played back.

Now you trim the audio so you can build a loop. Make sure that the Slip Mode edit-mode button is selected.

Figure 2-11
The drum track waveform

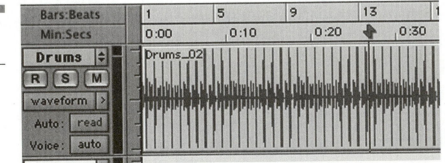

8. Using the Zoomer tool (the magnifying glass), zoom into the beginning of the first beat of your drum track. You may have to click several times.

9. Use the Trimmer tool to trim the audio region (clip) to just before the beginning of the first sound (see Figure 2-12).

10. Zoom back out until you can see the whole audio region.

11. Use the Operations > Loop Playback pull-down menu to select loop playback. You will see the Play button symbol in the Transport window has changed to the loop playback symbol.

12. Press the spacebar to start the loop playing. Listen to the beat and try to identify exactly where the end of one measure is.

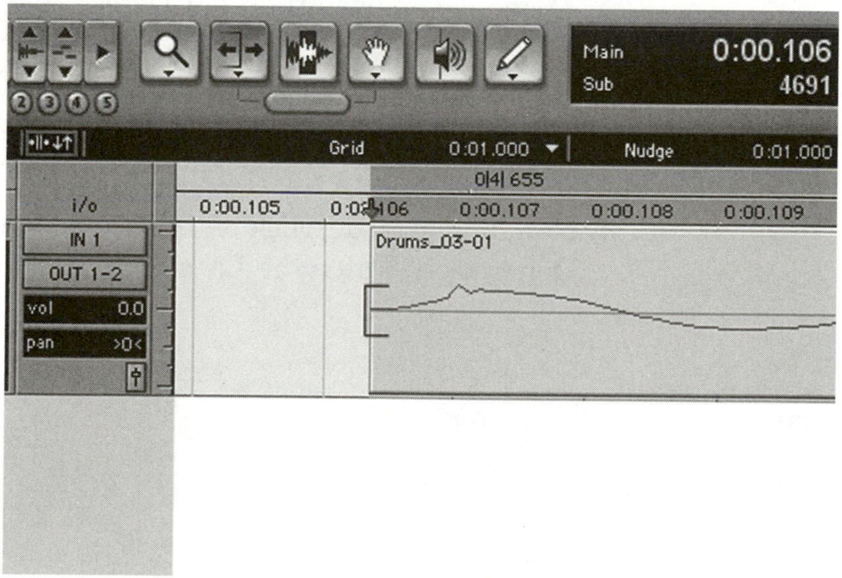

Figure 2-12
Using the
Trimmer tool

13. Use the Trimmer tool to trim as closely to that point as you can, so that what remains is exactly the length of four full beats. Loop-play the selection back again and see if it sounds like a steady beat. Trim the endpoint of the loop as necessary. Keep going back and forth between the Trimmer tool and loop playback until you are satisfied that you have a nice, steady, four-beat loop (or eight-beat). You should end up with one (or two) measure(s) of drums that will provide a consistent beat when looped.

14. Now, using the Grabber tool, drag the clip to the beginning of the timeline.

Beat Detective   The Beat Detective is a tool for extracting rhythmic information from an audio region that you can use to establish a grid of bars and beats that can be used to snap edits to.

1. Use the pull-down menu, Windows > Show Beat Detective, to open the Beat Detective window (see Figure 2-13).

2. Under Selection, enter the value of 1 for the start bar and also for the start beat. Enter the value 2 for the end bar and 1 for the end beat.

3. Now, under Mode, click on Bar/Beat Marker Generation. Watch the Bars:Beats ruler and you will

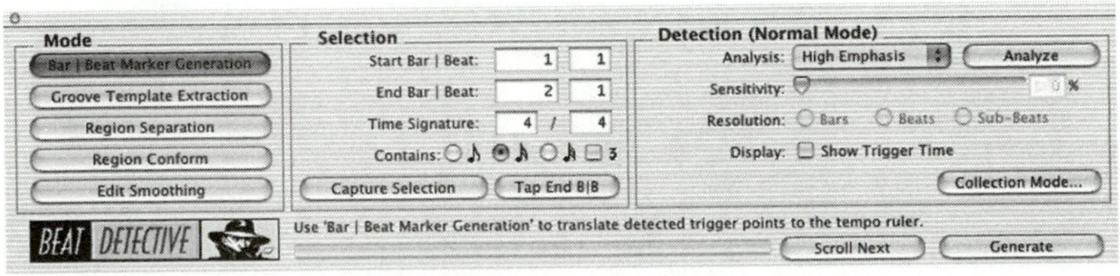

Figure 2-13
The Beat Detective window

see it adjust the time scale to match your selection. A
Bars:Beat grid will appear in the track that matches
the beat of your drum loop (see Figure 2-14).

Zoom Presets

Let's take a look at the zoom presets for a minute. You will
save yourself a lot of keystrokes and time by getting familiar with these tools and setting them up for your project.
They are at the upper left of the interface next to the edit
tools. See Figure 2-15.

1. The arrow at the right will zoom in and the arrow at
   the left will zoom out.

2. Use these tools to zoom out until the timeline is about
   a minute long. While holding down the command

Figure 2-14
The Bars:Beats
grid

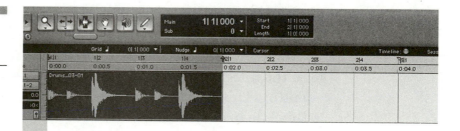

Figure 2-15
The zoom-preset
buttons

(control) key, click the Number 1 zoom preset button. From now on, if you click on that button or the control (start) key and the number 1, the timeline will zoom to this preset position.

3. Now, zoom in two clicks by holding down the command key and clicking the right bracket two times. Make this Zoom Preset 2 by holding down the command (control) key and clicking the Number 2 zoom preset button. Preset 2 is now set.

4. Continue in the same manner and set the remaining presets 3, 4, and 5, zooming in two more clicks for each preset.

5. Now, click the a . . . z symbol that is just under the zoom presets to activate the Commands Focus function. Now, if you press a number key between 1 and 5, the timeline will zoom to that preset.

Grid Mode: Copying, Pasting, and Consolidating Regions

1. In the upper left of the Edit window, select the Grid button. This edit mode will cause regions in the timeline to snap to the grid lines.

2. With the Grabber tool activated, click in the drum audio region to select it. Use command (control) + C to copy it.

3. Go to Zoom Level 1. With the Selection tool activated, click in the first open space in the timeline after the Drum region. (This will begin at Bar 2:Beat 1.) Hold down the shift key and click on the track about 1 minute down the timeline to the right. This will select

and highlight the part of the track between the two points.

4. Now, using the pull-down menu, select Edit > Repeat Paste to Fill Selection.

5. The Batch Fades dialog box will appear. Click OK to create very short crossfades and close the window.

6. Click anywhere on the drum track to deselect the clips. You will now see the same drum clip repeated in the timeline for about a minute. See Figure 2-16.

7. Return to the beginning of the timeline and press the spacebar to play to clips. It should sound like a repetitive drum loop.

8. Now, select all the drum clips in the timeline again. Go to the pull-down menu Edit > Consolidate Selection. This function will merge the clips into one solid drum track (see Figure 2-17).

   A new audio region and a new audio file are created by this operation.

9. Save.

Exercise 2-3: Overdubbing a Track

Overdubbing is the process of recording additional tracks to a session while hearing previously recorded tracks that are being simultaneously played back. This is the process that is at the heart of how most music is produced today. Before multitrack recording and overdubbing were invented, the only way to record a musical performance

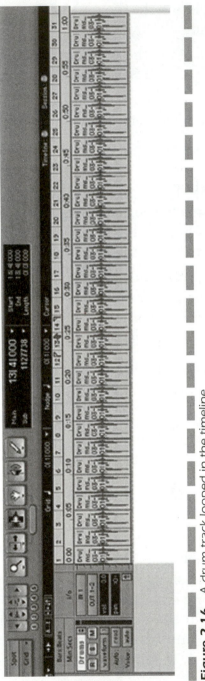

Figure 2-16 A drum track looped in the timeline

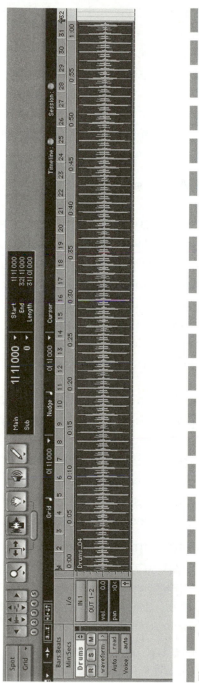

Figure 2-17 A consolidated selection

was to capture it live, with all the parts being played simultaneously.

In Pro Tools you can record one or more tracks simultaneously and build up your session one or more tracks at a time. Open whichever version of the drum recording session you prefer, tutorial_session_1 (original drum track) or tutorial_session_1a (looped drum track).

Monitor Modes

There are two monitor modes, and you will be switching back and forth between the two frequently during overdubbing. You use Input Monitor Only mode when you are setting up and rehearsing an overdub track and Auto Input Monitor mode when you are actually recording an overdub track. Since switching modes is a frequent operation, it is preferable to get used to using the keyboard-equivalent commands.

1. Use the pull-down menu, Operations > Input Monitor Only, or, preferably, the keyboard equivalent, command (control) + K. You will see that the Record button in the Transport window turns green. This color change means that you will be able to hear your new input instrument through the system while rehearsing as the previously recorded track(s) plays back.

2. Set up the instrument (or microphone) to be recorded by plugging it into Input 2 of your Pro Tools interface. If you are using a microphone, make sure your studio monitors are turned all the way down to prevent accidentally causing feedback.

3. The input to the second track (guitar, keyboard, and so on) should still be IN-2. (See the I/O view of the track.)

Put the track into record mode by clicking the R button.

4. Press the return (enter) key to move the playhead to the beginning of the track.

5. Press the spacebar to start playing the session.

6. Play your new instrument to check and adjust the input level so that you have a strong signal that is not clipping.

7. In the Mix window, adjust the playback level of the drum track, using the fader at the bottom, until you can comfortably hear the playback of both the drums and the new instrument through the headphones (or studio monitors if there are no live microphones).

8. Press the return (enter) key to move the playhead to the beginning of the track.

9. Use the pull-down menu, Operations > Auto Input Monitor, or the keyboard equivalent, command (control) + K. You will see the Record button in the Transport window turn back to its default gray color. This means that when you start recording you will hear the new input instrument as it is being recorded.

10. Press F12 to start recording.

11. Press the spacebar when you are done recording.

12. Take the track out of Record mode by clicking the R again so that is it no longer highlighted.

13. Hit the return (enter) key to move to the beginning of the session. Hit the spacebar to hear playback of the recorded tracks.

14. If you do not want to save the new track, use the command (control) + Z keyboard equivalent or the pull-down menu, File > Undo Record Audio.

15. If you want to keep the track, do a Save As and call
the session tutorial_session_1b.

You can record and save as many additional takes as
you want without having to create new tracks for each
take by using playlists.

Exercise 2-4: Playlists and the Playlist Selector

In the track, next to the track name, there is a small box
with up and down facing arrows. This is the playlist selector (see Figure 2-18).

1. Click the mouse and hold in the playlist selector and
select New (see Figure 2-19).

A dialog box will pop up asking you to name the new
playlist. Use the default and click OK.

Figure 2-18
The playlist
selector

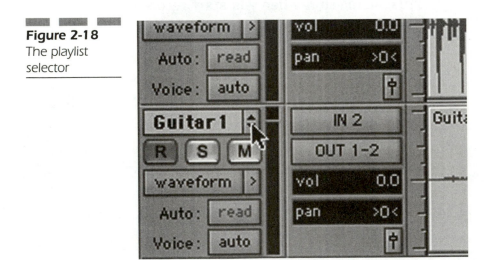

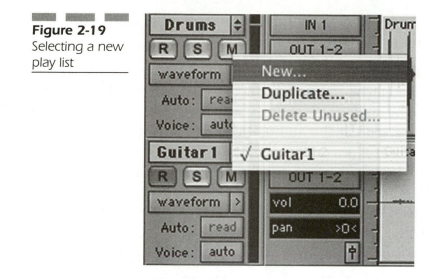

Figure 2-19
Selecting a new
play list

2. Press F12 to start recording.

3. Press the spacebar when you are done recording.

4. Repeat until you are satisfied with most of the track. Don't worry about a few mistakes. Those can be repaired later.

Renaming and Managing Audio Regions

On the far right of the Edit window is the audio regions list. Each time you do another take, the new material is added to the audio regions list. To rename audio regions, follow these steps:

1. Double-click on the last take you did. A Rename Selected dialog box pops up.

2. Type in a name that is more descriptive, such as "Final Guitar Take." Leave the name region and disk file selected so that both the audio region and the audio file will be renamed.

You will notice that some of the audio regions in the audio regions list are displayed in bold type. This means that the audio region represents the entire audio file. If an audio region in the audio file list is not bold, then it represents only a part of an audio file.

Look at the beginning of your "Final Guitar Take" audio region in the track. You will now see the new name displayed in the upper-left corner of the audio region. Rename the drum track the same way:

1. With the Grabber tool, select the audio region in the drum track. The track will also be selected in the audio regions list. Double-click on it and the Rename Selected dialog box pops up.

2. Type in a name that is more descriptive, such as "Final Drum Loop." Leave name region and disk file selected so that both the audio region and the audio file will be renamed.

3. Click OK to close the window.

Now, looking at the tracks in the Edit window, you should see both audio regions displaying their new names.

Deleting Playlists

If you keep creating playlists and doing takes, you will soon find that you have a lot of them that you are sure you are not going to use. You may want to discard these before you wind up with too much clutter in the audio regions list.

1. Select an unused audio region in the audio regions list.

2. Click and hold the mouse on the word *audio* at the top of the audio regions list to open the pull-down Audio Regions List menu. Choose "Clear Selected" (see Figure 2-20).

   This action will bring up the Clear Audio menu. See Figure 2-21.

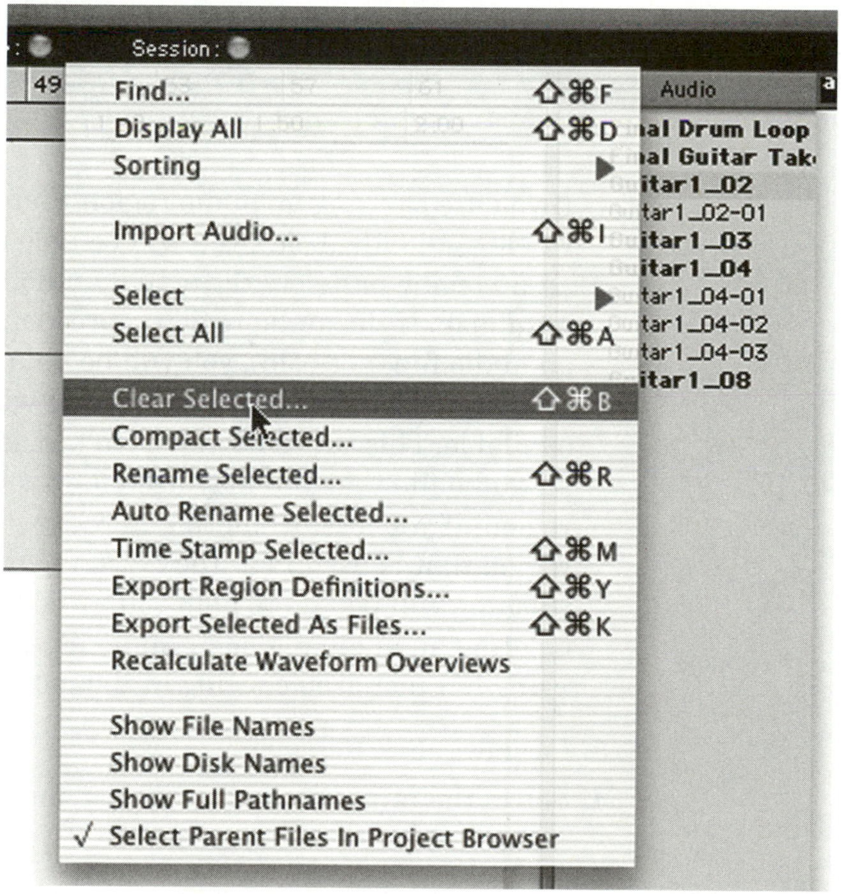

Figure 2-20
Clear selected in the Audio Regions menu

Figure 2-21
The Clear Audio
window

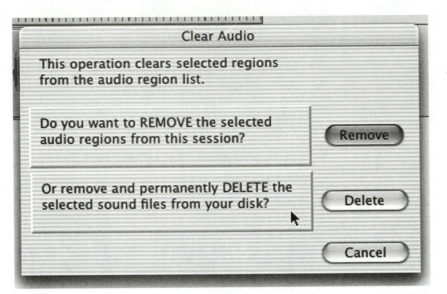

If you are clearing an audio region that represents a whole audio file (bold), you will be able to remove the audio region or delete the actual audio file from your hard disk. If the audio region represents only part of an audio file (not bold), you will be able to remove the audio region, but the Delete function will not be available. This prevents you from accidentally deleting an audio file that you may need for another audio region. If you are attempting to remove an audio region that is used in a track (even a hidden track), a warning will pop up to help you avoid accidentally deleting an audio region or audio file that you may be using but just aren't seeing at the moment.

3. For now, click Remove.

4. Select the other unused audio regions and delete them.

Figure 2-22
A sample audio
regions list

Your audio regions list should look something like
Figure 2-22.

Save your session by using the command (control) +
S keys.

Now use Save As and name the session tutorial_
session_1c.

Exercise 2-5: The Cue Mix

Now it's time to lay down a vocal or lead track. For this
track we will want to record with a microphone and we
must use headphones to hear the playback. Otherwise, the
drum and guitar tracks being played back over the studio
monitors will be picked up by the vocal microphone and
bleed into the vocal track. Even worse, it can cause feedback if the monitors are loud enough.

Pro Tools LE systems have dedicated headphone-output
jacks that allow the main mix to be routed through a pair
of headphones. You can control the headphone volume by
the available volume control on the front of the box. But
there is no way to output separate mixes to the headphone
out. So, if you are using an Mbox, technically there isn't

much point to the next exercise and you can just skip to Exercise 7 if you like. (However, you may want to go through exercise 6 anyway, just to get a feel for how the Sends and Aux tracks work. If you do, just use Output 1-2 where the exercise instructions say Output 3-4.)

With Digi 001, Digi 002, *time division multiplexing* (TDM), and HD systems, there are multiple outputs available that are addressable by routing the output of specific tracks directly to them or by using sends. This technique enables the use of separate mixes that can be routed to separate outputs. One of the uses of this capability is to create cue mixes.

In order to create the best performance, the vocalist will want to hear a cue mix in his or her headphones. This will be a mix that includes some or all of the recorded tracks mixed in with the live vocal microphone. Depending on all kinds of variables, the vocalist may want to hear more or less of the tracks or him- or herself. If there are multiple voices or instruments recording to different tracks simultaneously, he or she may want to have separate and different cue mixes.

In order to set up and route a cue mix, we will be introducing a few more features of the system. These include auxiliary (aux) tracks, sends, and busses.

Sends

The sends are always displayed in the Mix window. The Edit window can also be set up to display the sends.

1. Use the pull-down menu, Display > Edit Window Shows > Sends view, to display the Sends view in the Edit window. (I/O view should still be checked.) See Figure 2-23.

Figure 2-23
The Sends view
in the Edit
window

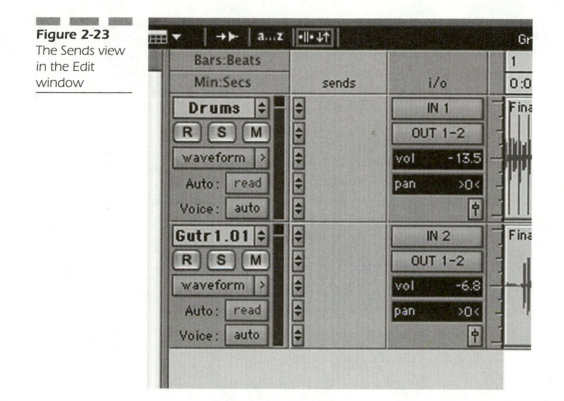

2. While holding down the option (alt) key, hold the
   mouse down on the top set of arrows in the Sends view
   in any track and drag to Bus > Bus 3-4 (Stereo) and
   release. (Holding down the option (alt) key is a short-
   cut that applies an operation to all visible tracks when
   it is performed on one track.) See Figure 2-24.

   This action will create a stereo send in all visible
   tracks. See Figure 2-25.

   A control strip with a fader automatically opens for the
   send. Close it by clicking the highlighted Bus 3-4 send
   in the track.

Figure 2-24
Creating a
stereo sound

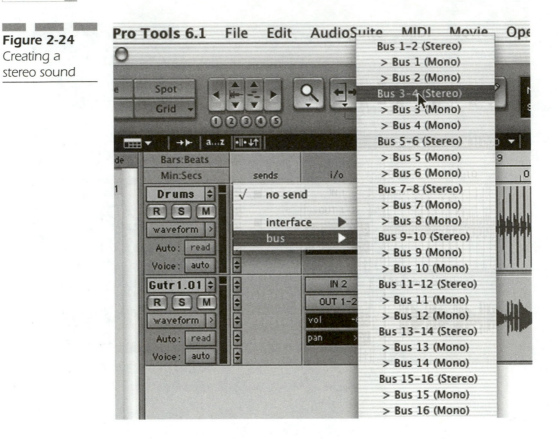

3. Use the pull-down menu, Display > Sends View
Shows > Send A. You will see minifaders, and other
controls appear in the Sends view. See Figure 2-26.

Auxiliary Tracks

Now, create a stereo aux input track, set the inputs to Bus
3-4, and set the outputs to Output 3-4:

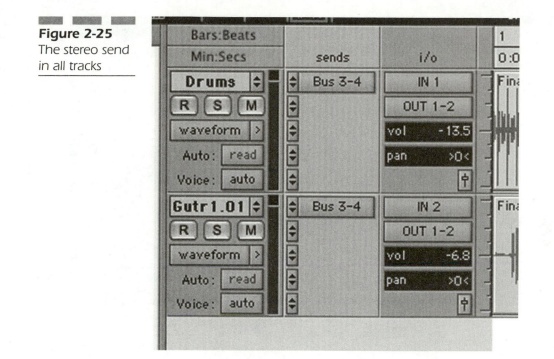

Figure 2-25
The stereo send
in all tracks

1. Open the New Track dialog box (File > New Track).
2. Use the pull-down menus in the dialog box to create one stereo aux input.
3. Rename the track "Cue Mix."
4. Set the input of the track to Bus 3-4. See Figure 2-27.

You should now connect Outputs 3 and 4 of your interface to your headphones, headphone amplifier, or an external mixer that can drive your headphones. (This will depend on your system configuration. If you are using an Mbox, just use the Outputs 1 and 2.)

Figure 2-26
New controls in
the Sends view

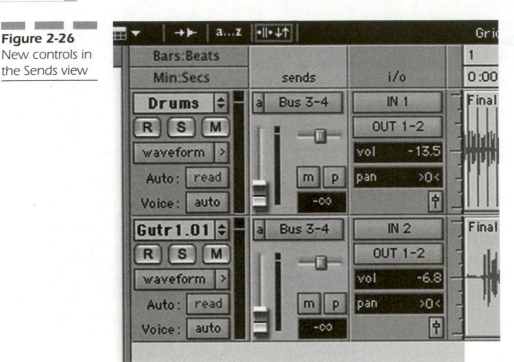

Figure 2-26
New controls in
the Sends view

Setting the Volume and Mix of the Cue Track

1. With the speakers turned all the way down, start playing the session back from the beginning.

2. While listening to the headphones, move the mini-fader in the drum track up until you have a comfortable level.

3. Do the same with the second track. You should be able to hear both tracks through the headphones. Next to the minifader there is a pan control. Play with that a bit and see how you can pan each track independently. For now, pan both tracks all the way left.

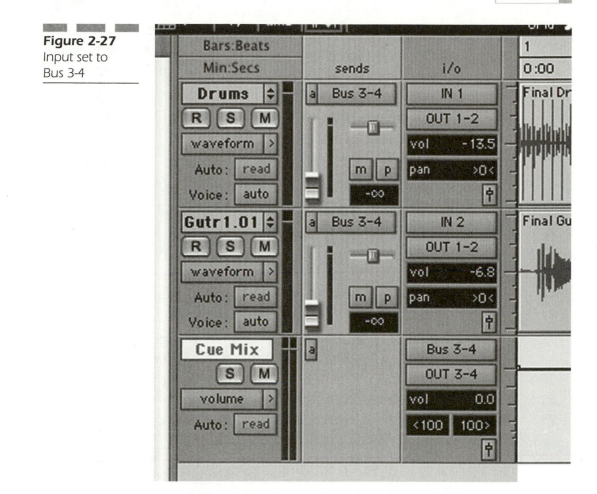

Figure 2-27
Input set to
Bus 3-4

Setting Up the Vocal Track for Recording

1. In the Show/Hide column at the left, click on the Vocal
1 track to unhide it.

2. Make sure that the microphone is on and that the
input of the Vocal 1 track is set to whatever input of
your interface the microphone is connected to.

3. Make sure that the monitor mode is set to Input Only Monitor (Operations > Input Only Monitor). When you sing or play into the microphone, you should see a signal in the track's meter.

4. Put the track into record by clicking the R button. (It turns red.)

5. Sing or play into the mic and adjust its level for a strong input signal that is not clipping.

6. In the Sends view of the Vocal 1 track, route the track's send to Bus 3-4.

7. Pan the send all the way to the right. When you sing or play into the microphone, you should see the stereo meter in the cue-mix track indicate that the signal is going to the right stereo channel only.

8. Now play your track back from the beginning and sing. You should hear the previously recorded tracks in the left headphone and the vocal in the right headphone. See Figure 2-28.

9. Adjust the pan and volume of the sends to taste.

You are now ready to record the vocal track. Save your session and then do a Save As with the session name tutorial_session_1d.

Exercise 2-6: Recording the Vocal Track Without a Cue Mix

If you are using an Mbox system and skipped Exercise 6, or if you are not going to use a cue mix (and skipped Exer-

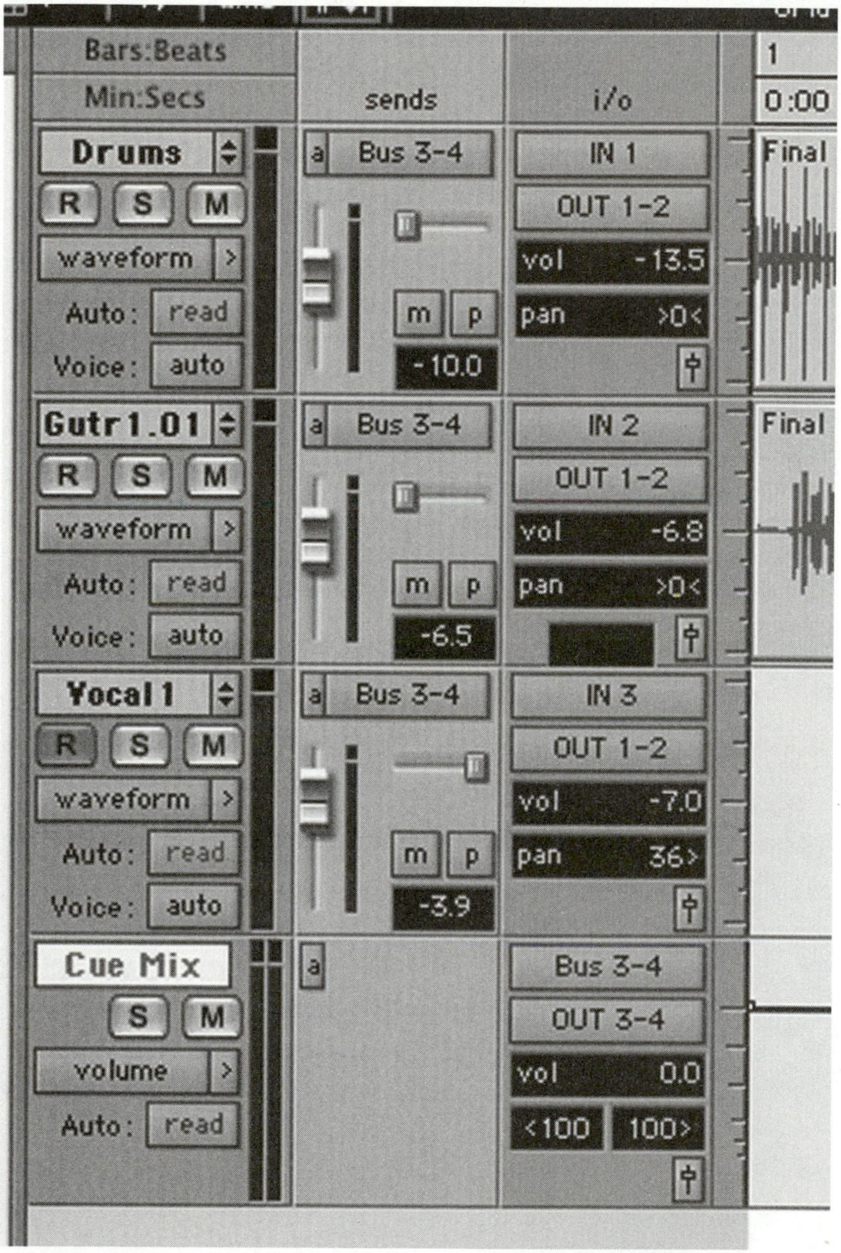

Figure 2-28 Play back the track while recording

cise 6), there are a few steps you need to take to prepare to
record the Vocal 1 track:

1. In the Show/Hide column at the left, click on the
   Vocal 1 track to unhide it.

2. Make sure the studio monitors are turned down.

3. Make sure that the microphone is on and that the
   input of the Vocal 1 track is set to whatever input of
   your interface the microphone is connected to.

4. Make sure that the monitor mode is set to Input Only
   Monitor (Operations > Input Only Monitor). When you
   sing or play into the microphone, you should see a
   signal in the track's meter.

5. Put the track into record by clicking the R button. (It
   turns red.)

6. Sing into the mic and adjust its level for a strong input
   signal that is not clipping.

7. In the Mix window, using the pan sliders in each track,
   pan the prerecorded tracks all the way to the left.

8. Pan the Vocal 1 track all the way to the right.

9. Now play your track back from the beginning and sing
   or play. You should hear the previously recorded tracks
   in the left headphone and the vocal in the right
   headphone. Use the volume sliders in the Mix window
   to adjust the playback levels and the panning position
   of the tracks until you have the mix you want to use
   for recording.

Exercise 2-7: Recording the Vocal or Solo

1. You can use the pull-down menu, Operations > Auto Input Monitor, to set the monitor mode to Auto Input Monitor. However, you should start using option (alt) + K on the keyboard to toggle between Input Monitor Only and Auto Input Monitor modes. You will be going back and forth between these modes all the time when you are overdubbing and it's much easier to use the keyboard. When you are in Input Monitor mode, the Record button in the Transport window turns green. In this mode, you will be able to rehearse and hear the live input to tracks that are armed to record (their R is red) through the system as you rehearse. When you are ready to record, toggle back to Auto Input Monitor. In this mode, you will hear what you are recording to the new tracks as you do it.

2. Use the return (enter) key to move to the beginning of the track.

3. Hit the F12 key to start recording.

4. When you are done recording, press the spacebar to stop.

5. Take the track out of record.

6. Return to the beginning and playback the session. You should now hear all the recorded tracks.

7. Rename the Vocal 1 audio region Vocal_1_Final.

8. Save your session and then do a Save As with the name tutorial_session_1e.

Exercise 2-8: Punching In

As you listen to your tracks, you may decide that you like most of the vocal, but feel you would like to rerecord a few bars. So, you decide that you want to punch in an eight-bar segment of the vocal track. Actually, you want to do it over and over without stopping until you get one you like. But you want to save all of the takes just in case. Fortunately, this is one of the many things that Pro Tools is really good at.

Loop Record, Pre-Roll, and Post-Roll

1. While holding down the control (start) key, click several times on the Record button in the Transport window. It will toggle through the four record modes. The default mode is nondestructive recording. In this mode, the Record button appears normal. In Destructive Record mode, there is a small D in the center of the button. In Quickpunch mode, there is a small P in the center of the button. In Loop Record mode, there is an arrow looping around the button. These controls are also available under the Operations pull-down menu. Set the record mode to Loop Record (see Figure 2-29).

2. Make sure your edit mode is still set to grid. With the Selector tool, drag across the section of the Vocal 1 track that you want to punch in on (see Figure 2-30).

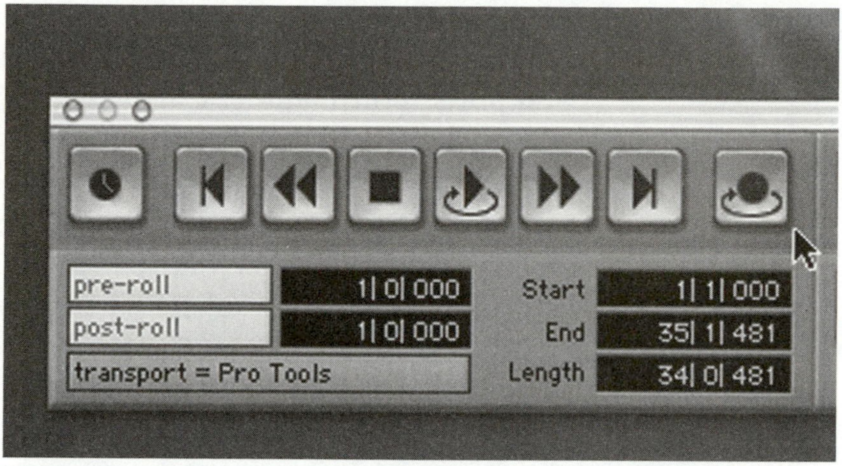

Figure 2-29
The record mode set to loop

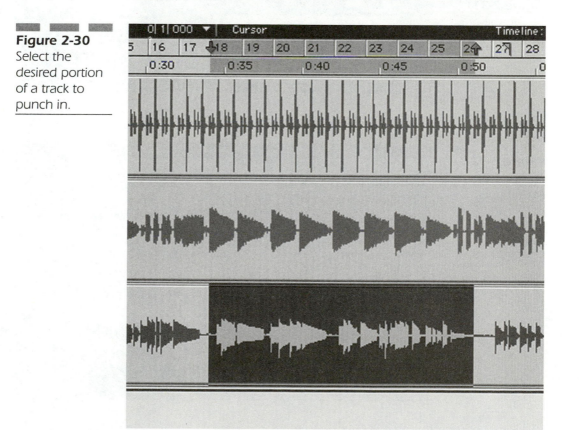

Figure 2-30
Select the desired portion of a track to punch in.

3. In the Transport window, click the 1 in the number
   field next to the Pre-roll button. Enter the value *6*
   and press return (enter). This will give you a six-bar
   pre-roll before the punch point. Leave the post-roll at
   the default of 1 bar (see Figure 2-31).

   Look at the Bars:Beats ruler. The red arrows indicate
   the in and out points for your punch in. The green
   flags indicate the pre-roll and post-roll points. See
   Figure 2-32.

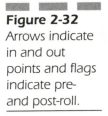

Figure 2-31
Setting the pre-
and post-roll

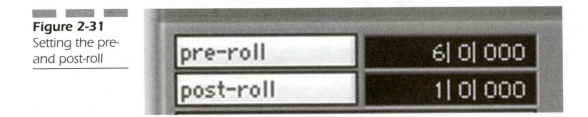

Figure 2-32
Arrows indicate
in and out
points and flags
indicate pre-
and post-roll.

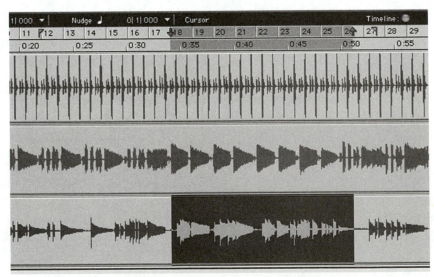

4. If you want to adjust any of these points further, you can do so by dragging the arrows or flags with the Grabber tool.

5. Make sure that the microphone is still ready to record with the same levels as before. Put the Vocal 1 track into record mode by clicking the R button. It will turn red.

6. Hit the F12 key. The pre-roll will start. When it gets to the in point, recording will begin. You can listen to the existing tracks as the pre-roll plays so that you are ready when it gets to the in point and starts recording. In Loop Record mode, as soon as the playhead gets to the out point it will jump back to the in point and keep recording. Go ahead and keep singing or playing for five or six complete cycles. Hit the spacebar to stop recording.

7. Take the track out of record mode. If you look at your audio regions list, you will see a new playlist for each of the five or six takes. See Figure 2-33.

The new tracks have all been recorded and saved with names generated by the system. The program has also created two new regions, Vocal_1_Final_01 and Vocal_1_Final_02, that represent the parts of the original track that come before and after the punch-in. There is a new audio region in the track and it is the last take that you input. You can now play back the session and when it gets to the new audio region, the program will play the new version instead of the previously recorded part. If you want to hear the other takes, you can drag them from the audio region list into the track and place them directly over the audio region that you are working to replace. The

The audio regions list showing:

Audio   a
                               z

Final Drum Loop
Final Guitar Take
Vocal1_08
Vocal1_08-01
Vocal1_08-02
Vocal1_08-03
Vocal1_08-04
Vocal1_08-05
Vocal1_08-06
Vocal_1_Final
Vocal_1_Final-01
Vocal_1_Final-02

entire process is nondestructive, so you can always go back to any one of the previous takes, including the original. So, try dropping the different takes you did into the track and keep the one you like the best. If you don't like any of them, you can go back and try again. When you get the track the way you want it, save the session.

Exercise 2-9: Comping a Track

It is possible that you liked parts of a couple different punch-ins. Let's assume that you want the first half of the first take and the second half of the second take. You can build a comp track and use the parts you like from each one:

1. Hide the cue mix track.
2. Make sure the first take of the punch-in is in the Vocal 1 track.

3. Create a new track and call it Vocal 2.

4. Drag the audio region for the second take from the audio regions list into the Vocal 2 track so that it snaps to the grid and is directly under the first take in the Vocal 1 track.

5. In the Vocal 1 track, use the Trimmer tool to trim back from the end of first take's audio region until only the first half remains. There should now be a gap in the track.

6. In the Vocal 2 track, use the Trimmer tool to trim from the beginning of the audio region so only the second half remains and the beginning of the audio region is just below the end of the audio region in the Vocal 1 track.

7. Create another new track and call it Vocal Comp.

8. Drag all the clips in the Vocal 1 and Vocal 2 tracks down into the Vocal Comp track.

9. Mute the Vocal 1 and Vocal 2 tracks.

10. Play the session and you should now hear the Vocal Comp track with your new selection of audio regions.

You can now add and edit additional tracks and record more audio as needed using the same procedures as above. As noted at the beginning and throughout the book, there are always going to be alternative methods and procedures for accomplishing almost any task in Pro Tools and there are many recording features in addition to those described in these exercises.

CHAPTER 3

Mixing

About Recording, Editing, and Mixing

In the recording process, audio is turned into digital information and stored on hard drives as audio files in an audio file format. When the recording takes place, the system also generates an audio region file that is a graphical abstraction of the audio file.

In the editing process, the audio regions are moved, cut, copied, pasted, duplicated, and otherwise manipulated in a graphical interface. The result is an instruction set, stored in the Pro Tools session file and expressed in the graphical timeline of the Edit window that defines the playback selection and timing of the audio files.

The mixing process adds additional instructions to the Pro Tools session file and they control the digital-signal processing and routing that will be applied to the audio files when they are played back.

In Pro Tools, the actual process of recording and playing back the audio files is a computer-based process of creating, storing, and retrieving data. It has very little in common with the way audio is recorded and played back in an analog system with tape machines and mixing consoles. However, the graphical interface is set up to resemble an analog recording environment for purposes of giving the operator a familiar and understandable working environment.

Exercise 3-1: Basic Mixing

Load a session into Pro Tools. If you have followed the exercises up until now, use the session you have been working on.

1. Use command (control) + N to go to the New Track dialog. Create a new Stereo Master Fader track.

2. Use the large faders in the Mix window and bring the volume of all the tracks except the Master Fader down to about -7.0. These faders control the output volume. See Figure 3-1.

Figure 3-1
Output volume controls in the Mix window

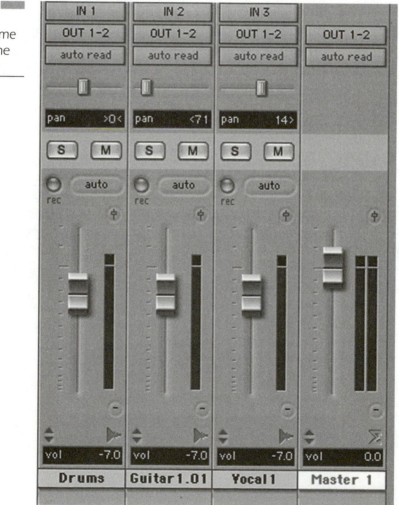

3. Play the session from the beginning.

4. Use the faders to adjust the volume up or down on each track as needed to achieve a good blend of volume. Notice how the changes affect the meters in the tracks and in the master fader. Avoid clipping in all tracks.

5. Using the pan faders in the tracks in the Mix window, pan the second (guitar, instrument) track to the left about halfway and the vocal (lead) track to the right. Listen to the panning effect and position the tracks in different positions. See Figure 3-2.

6. Use the master fader to set the optimal output volume and avoid clipping.

7. Save the session.

Plug-ins and Auxiliary Sends

Mixing in Pro Tools extends the analog analogy of the graphical interface to include not only a tape machine and

Figure 3-2
Adjusting the
pan of a track

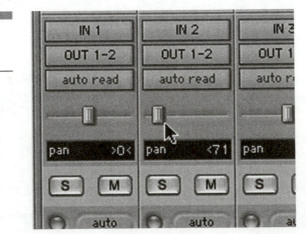

a mixing console, but additional outboard equipment as well. In an analog recording studio, various outboard hardware devices are connected by physical wires to patchbays or routing switchers. In Pro Tools, plug-in software takes the place of the outboard devices, and the use of inserts, sends, and virtual busses takes the place of the wiring.

The following exercises demonstrate how they are applied and some basic considerations in plug-in use. There are many types and variations of plug-ins. There is a lot more information on the specific plug-in types and functions later in the book.

Exercise 3-2: Using AudioSuite Plug-ins

1. Use the Selection tool to select a 5- to 10-second part of an audio region in a vocal track.

2. Use the pull-down menu, AudioSuite > D-Verb, to bring up the D-Verb control panel.

3. Use the Input slider to bring the input level up to 0.0 *decibels* (dB). See Figure 3-3.

4. Press the Process button located at the lower right of the control panel. You will see a Processing window pop up for a few seconds and then disappear. A new audio region will appear in the track and in the audio region list. Its newly generated name will contain elements of the name of the original file and the name of the plug-in.

5. Play the section of the track to hear the new effect.

6. If you want to alter the settings for the plug-in, use command (control) + Z or the pull-down menu, Edit

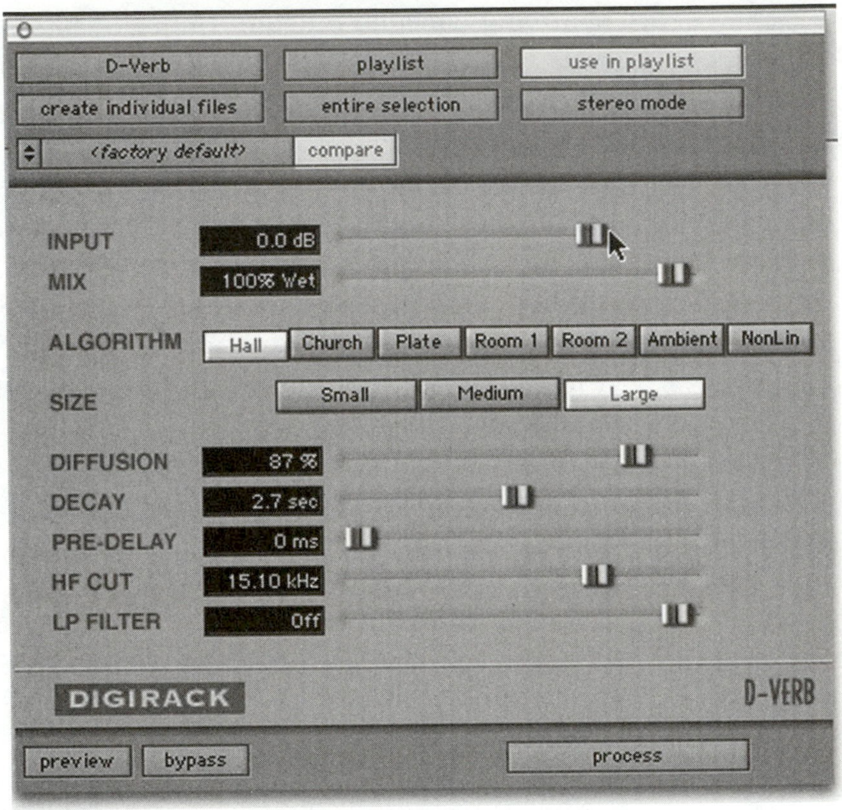

Figure 3-3
Using the Input
slider

> Undo D-Verb, to undo the plug-in. Repeat steps 1 through 5 with new settings.

There are many different kinds of plug-ins. They each have their own control panel that will pop up when opened. The control panels will differ and be specific to the functions of the particular plug-in.

Plug-in Architectures

Pro Tools currently supports three different kinds of plug-in architectures that operate in different ways within the

Pro Tools system (*Hybrid time division multiplexing* [HTDM] plug-ins are actually a fourth type that will be discussed later):

■ AudioSuite plug-ins are not real-time, but file based. The effect of an AudioSuite plug-in cannot be heard until it is rendered or processed. In AudioSuite plug-in control panels, there is a control called Process. After setting up the parameters in the control panel, pressing the Process button will cause the system to render a new audio file that is the original audio with the plug-in's process applied.

■ *Real-time audio suite* (RTAS) plug-ins are, as their name suggests, real time. This means that you can make settings as you are playing audio and hear the effect of the changes immediately. RTAS plug-ins use your host computer's *central processing unit* (CPU) to power the effects.

■ TDM and TDMII plug-ins are also real time. However, they require specialized Digidesign hardware, specifically Mix Core or Mix Farm cards or the HD Core or HD Process cards (required for TDMII plug-ins).

The next exercise uses another version of the D-Verb plug-in. If you have a TDM system, use the TDM version of the plug-in. If not, use the RTAS version.

Exercise 3-3: Using RTAS and TDM Plug-in Inserts

1. If you haven't done so already, use the pull-down menu Display > Sends View Shows > Assignments.

2. At the top of the vocal (lead) channel strip in the Mix window, hold the mouse down on the top set of up and down arrows. These are the insert selectors. A pull-down menu will appear. Select RTAS Plug-in > D-Verb (mono) (see Figure 3-4).

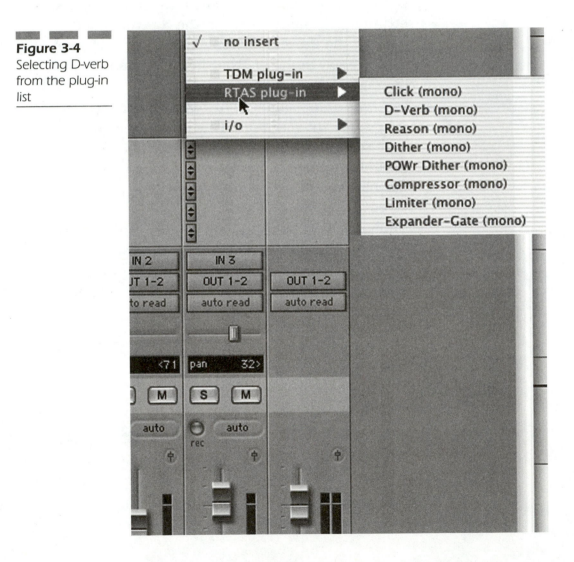

Figure 3-4
Selecting D-verb from the plug-in list

If you are using a TDM system, you could also select
the TDM version of the plug-in in the same manner,
TDM Plug-in > D-Verb (mono).

3. The D-Verb control panel will pop up. Use the input
slider to bring the input level up to 0.0 dB. See
Figure 3-5.

4. While playing back the session from the beginning,
move the mix slider throughout its range and listen.
This determines the ratio of output signal that is
unprocessed (dry) and processed (wet) by the plug-in.

Figure 3-5
Adjusting input
in the D-Verb
control panel

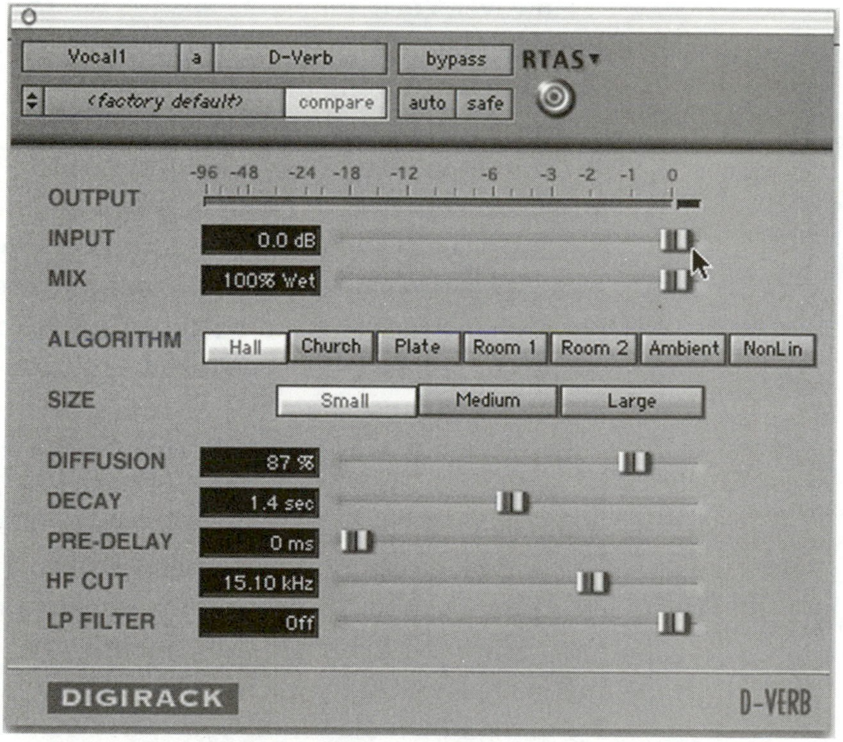

5. Try the different algorithm presets. Notice how they affect the other individual controls and how they affect the sound.

6. When you are done exploring the plug-in, delete it by selecting "no send" again in the track.

Conserving Processing Resources

When you use an AudioSuite plug-in, you are creating a new audio file that is based on the original audio file with the changes effected by applying the plug-in. When you play it back, you are simply playing back a new audio file. This use places no additional requirements on your system than playing any other audio file.

However, each time you use an RTAS plug-in, you are using up some of your host computer system's available real-time processing power because the effect processing is taking place in real time. Similarly, each time you use a TDM plug-in, you are using up some of the available *digital signal processing* (DSP) processing power on the TDM cards in your system. There is a limited supply of these processing resources. As you keep adding more RTAS or TDM plug-ins to a session, you will eventually run out of system power to support all those real-time processes. So, it is advisable to conserve processing resources where possible.

One obvious way to conserve is to use AudioSuite plug-ins when possible. Many plug-ins, such as D-Verb, come in AudioSuite format, as well as RTAS or TDM.

Another way to conserve resources is to use one instance of a plug-in to do processing for multiple audio tracks. In the previous exercise, by setting up D-Verb in the instructed manner, you made it available to only one track. By using Pro Tools sends, you can route (send) the signals from multiple tracks over a bus to an auxiliary (aux) track. You can then insert the plug-in on the aux track, thereby using only one instance of the plug-in but applying the plug-in effect to the signal from multiple tracks.

This technique is very similar to what must be done in an analog studio that has limited outboard effects processors. For instance, if you wanted to apply a reverb effect from a special outboard reverb unit to four tracks, but you only had one of the units, you would route the insert signals of the four tracks to the reverb unit by physical patches over a physical bus. The output of the reverb unit would be sent back into the main mix and mixed in with the original signals. The level of reverb applied to each track would be determined by the amount of signal sent to the reverb unit from each track. The next exercise demonstrates how the digital version of this operation is done in Pro Tools.

Exercise 3-4: Using Sends and Aux Tracks for Plug-ins

1. Make sure that there are no plug-ins in use. Use undo to get rid of any AudioSuite plug-in effects and make sure that all the inserts and sends are in the no send or no insert state. Hide any master tracks or cue mix tracks you may have created before by unclicking them in the Show/Hide column on the left of the Edit window.

2. Using the New Track dialog, create one stereo, aux-input track.

3. Name the track D-Verb by double-clicking on its name and entering D-Verb in the Name the Track field.

4. Set the input of the D-Verb track to Bus 1-2. Leave the output set to Output 1-2.

5. Using the insert selector pull-down menu in the track, select multichannel plug-in > D-Verb (stereo). This action will insert the plug-in and open the D-Verb control panel.

6. In the Mix window, click on the first (top) send of each audio track and select Bus 1-2. The Send window will appear for each track as you do this. They each have a slider that can be used to control the amount of that track's signal you send to the bus (or, in this case, to the D-Verb). But just close those Send windows for now. We will use another set of faders instead. The top half of your Mix window should look something like Figure 3-6.

7. Use the pull-down menu, Display > Send View Shows > Send A. The Send controls with minifaders will appear in the sends area of each audio track in the Mix window. See Figure 3-7.

8. Mute all the audio tracks except the first one and the D-Verb track. Play the selection from the beginning and listen as you raise the minifader in that track until you hear the reverb effect. The reverb effect is coming from the D-Verb channel.

9. Mute the first audio channel and unmute the second one. Play back the session and adjust the send mini-fader for each track until you hear the amount of reverb you want for each individual track.

10. Unmute all the tracks and play the session. You can further adjust the D-Verb parameters and the send volumes as needed on the fly.

Figure 3-6
The Mix window

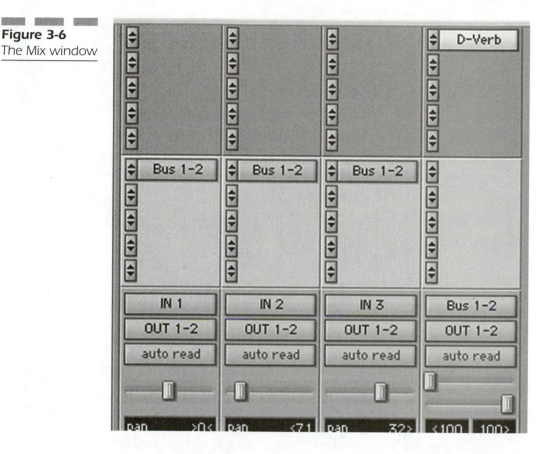

There are also multimono versions of many plug-ins, including D-Verb. These versions allow you to create two sets of settings while using only one plug-in.

11. Click on the input selector in the D-Verb track where you have the D-Verb plug-in selected. Select Multi-mono plug-in > D-Verb to replace the previously selected multichannel version.

12. Center the pan position on the two channels in the aux track by option-clicking them. See Figure 3-8.

Figure 3-7
Sends controls
in the Mix
window

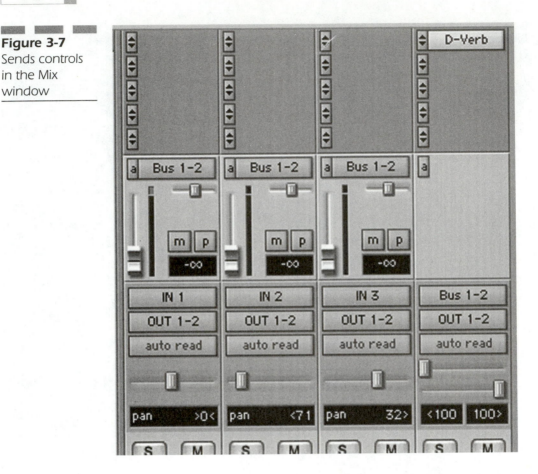

Figure 3-7
Sends controls
in the Mix
window

13. In the upper-right corner of the D-Verb control panel, click the highlighted Master Link button to unlink the two channels of the multimono D-Verb. See Figure 3-9.

14. Next to the Master Link button is the Channel Selector button. It can be set to L for left or R for right. Select L.

15. Choose Room 1 in the algorithms. Choose small for size. This is a short reverb.

16. Using the channel selector, select the R channel.

Figure 3-8
Centering the
pan position

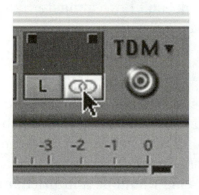

Figure 3-9
The Master Link
button

17. Choose the Hall algorithm and the large size. This is a longer reverb.

18. In the first audio track, change the send to Bus 1 mono.

19. In the other audio tracks, change the send to Bus 2 mono.

You now have two separate reverb channels programmed in the multimono D-Verb. The Bus 1 sends are going to the L channel and the Bus 2 sends are going to the R channel.

Exercise 3-5: Recording to Tracks/Submix

Recording to tracks is a method of creating a submix within Pro Tools. To do so, follow these steps:

1. Load the session that you have been working on in the previous exercises. Make sure all tracks are visible. Hide or delete any cue mix tracks. Do any additional mixing and add whatever plug-ins you want to use until you feel you have an acceptable mix.

2. If your session doesn't have a master fader, create a stereo master-fader track.

3. Set the output of all tracks to Bus 5-6 (stereo). (If this bus is in use, use an open stereo bus.)

4. Create a new stereo audio track and name it Submix. This will be the stereo track that everything gets mixed down to. If you were creating a mono mix, this would be a mono track. If you were creating a surround mix, this would be a multichannel track that matched the final surround format.

5. Set the input of the submix track to Bus 5-6 (stereo). The output should be set to your main audio outputs. Put the track in record mode by clicking the R. At this point, the upper part of the Mix window should look something like Figure 3-10.

6. Make sure the monitor mode is set to Input Only Monitor.

7. Play the session from the beginning. You should hear it through the speakers.

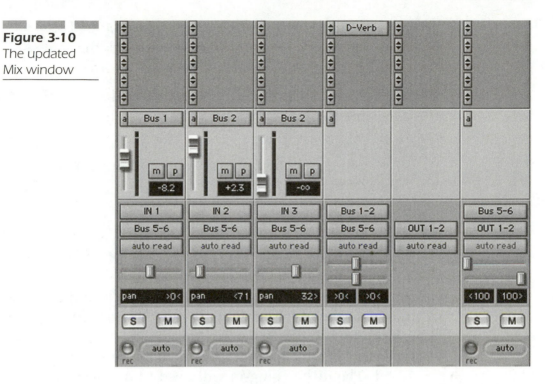

Figure 3-10
The updated
Mix window

8. While listening to the session and watching the faders, use the master fader to optimize the volume. The signal should be strong but not clipping.

9. If you are not recording the entire session, use the selector tool to select the portion that will be recorded. If you are recording the whole session, hit return (enter) to move the playhead to the start of the session.

10. Set the monitor mode to Auto Input Monitor.

11. Hit F12 to begin recording. You will see the submix track being recorded.

12. If you are recording a selection, recording will stop at the end. If you are recording open-ended, then press the spacebar when enough has been recorded.

13. Take the submix track out of record and solo it (S button). Play it back to hear the new submix.

14. The new Submix file now exists in the Audio Files folder of the session. It can be used in any application that accepts the format.

Exercise 3-6: Bouncing to Disk

Bouncing to disk creates a new audio file from the session in a wide variety of formats. In our ongoing example, the session we have been working on has been done in 24-bit. In order to burn an audio CD, it must be reformatted to 16-bit. The dither plug-in will need to be inserted on the master fader to accomplish this conversion. The result will be a 16-bit file. Even if you have been working in 16-bit throughout the session, using a dither plug-in can improve the final sound quality. Bouncing to disk can be done by following these steps:

1. In the Master 1 track, use the insert selector to add the POWr Dither (stereo) insert to the track. Choose 16-bit as the bit resolution and leave Noise Shaping Type 1 selected.

2. Use the pull-down menu, File > Bounce to Disk, to open the Bounce window.

3. Set the bounce options to Bus 5-6 (Stereo) (or whatever bus your tracks are routed to), BWF (.WAV), Stereo interlaced, resolution 16, sample rate 44100, and Convert After Bounce. See Figure 3-11.

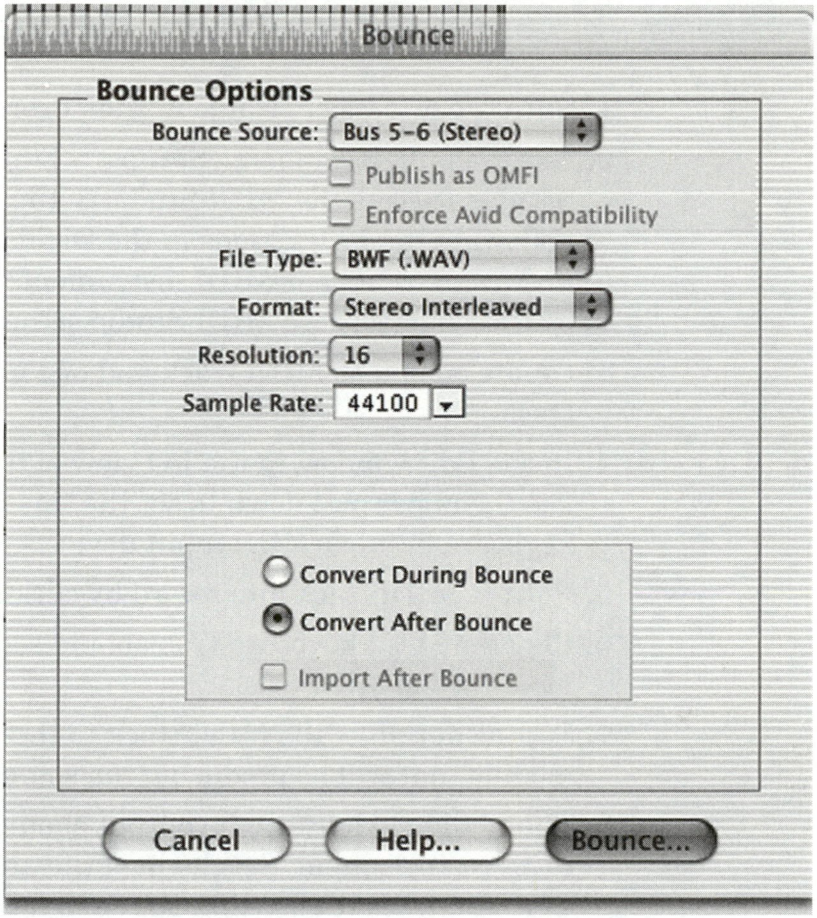

Figure 3-11
The Bounce
Options
window

Bounce

Bounce Options

Bounce Source: [Bus 5-6 (Stereo) ⬍]

☐ Publish as OMFI

☐ Enforce Avid Compatibility

File Type: [BWF (.WAV) ⬍]

Format: [Stereo Interleaved ⬍]

Resolution: [16 ⬍]

Sample Rate: [44100 ▾]

○ **Convert During Bounce**

◉ **Convert After Bounce**

☐ **Import After Bounce**

(Cancel) (Help...) (**Bounce...**)

4. Press the Bounce button.

5. When the Save As dialog appears, name the session "16-bit version," and click Save.

6. The session will play through. The 16-bit mix will be saved in your Audio Files folder.

7. If you want to, you can take the new 16-bit session file and import it into a CD-burning program to make a CD.

Exercise 3-7: MIDI Recording

1. Start a new session. Name it Midi_exercise_session_1.

2. Use the pull-down menu, Setups > Edit MIDI Studio Setup, to access the MIDI Studio Setup menu. Following the instructions in the Digidesign manual, make sure that your MIDI instruments are recognized by the system. Close MIDI Studio setup.

3. Create one mono MIDI track and one mono audio track.

4. In the Edit window, in the I/O view of the MIDI track, use the upper pull-down to set the "in" to your MIDI keyboard or other MIDI input device.

5. Put the MIDI track into record by clicking R.

6. Play the MIDI keyboard. You should see activity in the meter in the MIDI track.

7. Set the output of the MIDI track to the MIDI sound module that will be producing the sound.

8. The sound module's audio output should be connected to a Pro Tools input. Set the in of the audio track to that input. Make sure that the output of the audio track is set to your main audio outs.

9. Put the audio track into record mode. (Figure 3-12 shows how the controls in the two tracks should look in the Edit window. In this case, the MIDI controller is "Oxygen-1," and the sound module is "TrinityPrX-6." The sound module is connected to IN 4.)

10. When you play the MIDI keyboard, you should see the meters in both tracks move and hear the sound from the audio track. Adjust the volume of the sound module if necessary.

Figure 3-12
Controls in the
Edit window

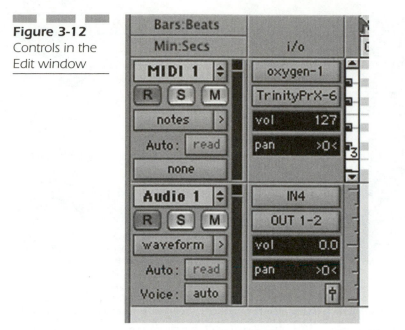

11. Hit the F12 key and start recording. In this case you
are recording both the MIDI track (generated by your
MIDI controller) and the audio track from the audio
output of the MIDI module. See Figure 3-13.

Exercise 3-8: Recording a Virtual Instrument with ReWire

Virtual instruments offer many advantages over external
sound modules and are becoming increasingly popular.
Pro Tools 6.1 offers support for ReWire (from Propellerheads Software). For this exercise, we will use the application "Reason," which is a software-based virtual music

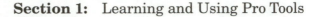

Figure 3-13
Recording a
MIDI track

studio application with various synthesizers and drum
machines included. If you don't have it, you can download
a working demo version of it from the Propellerheads Web
site at www.propellerheads.se. To record a virtual instrument using ReWire, follow these steps:

1. Start a new session and create two stereo audio
   tracks.

2. Use the plug-in selector in the Audio 1 track to select
   Multi-Channel RTAS Plug-In > Reason (Stereo). Wait
   while the Reason program automatically launches.

3. Select a demo song in Reason.

4. The Rewire/Reason control panel popps up when you
   activate the plug-in. Set the Reason output to Mix L -
   Mix R. See Figure 3-14.

5. Click on the Pro Tools Edit window to select Pro Tools.
   Hit the spacebar. The playhead will start moving and
   Reason will play. The meters in the Audio 1 track show
   activity.

Figure 3-14
Seting the
Reason input

6. Set the output of the Audio 1 track to Bus 1-2 (stereo).

7. Set the input of the Audio 2 track to Bus 1-2 (stereo).

8. Put the Audio 2 track into record by clicking the R in the track. See Figure 3-15.

9. Press the F12 key. Reason will start to play and recording will commence on the Audio 2 track.

Figure 3-15
The Audio 2
track in record
mode

2

The
Extended
Software
Environment

Building a Pro Tools System

Pro Tools is not just a program anymore. It has evolved into a platform that is supported by hundreds of additional programs from Digidesign and dozens of other companies. This extended software environment has developed into a rich virtual studio space that encompasses software solutions that emulate many kinds of physical devices, as well as many completely new digital tools.

These programs come in the form of plug-ins that work within the Pro Tools software environment, stand-alone software that works on the Pro Tools hardware platform, and software and hardware that are used on the same computer platform to enhance the overall working environment. The following chapters provide descriptions of a wide range of software that can be used to expand Pro Tools.

There are many companies involved in making Pro Tools-compatible software, and new versions of software are constantly being introduced. Most software is now being written to be compatible with Macintosh System OS X and Windows XP, which are the operating systems for Pro Tools 6.1 and beyond. As of this writing, some of the products are still in transition to their latest formats and new versions are still being introduced. The web sites for all products are listed with their descriptions and are also on the included disc. The www.ptbible.com web site will also be a resource for the latest information. The reader is encouraged to check these web sites for the latest details.

About Plug-ins

A plug-in is a software program that generally connects to a more complex program to enhance the larger program's functions or add capabilities. The concept of software plug-ins originated with graphic creation software like Adobe Photoshop and QuarkXpress, which use software developed by third parties to add specific capabilities to the main program.

A group of enterprising companies working with Digidesign brought this concept to Pro Tools. After years of development, many plug-ins are available that serve to extend Pro Tools' capabilities in many ways.

There is a certain logic to these software developments. Pro Tools itself replaces the main hardware components of an audio recording studio, including the multitrack tape recorder and the mixing console. However, there are still many outboard devices for signal control, manipulation, and effects that were not included in the basic Pro Tools software. So, the plug-in developers started building software emulations of these outboard devices that could perform the same functions within Pro Tools by plugging in.

Effects and signal-processing plug-ins are still the biggest part of the plug-in universe. However, over the last few years there has been a serious migration from hardware-based synthesizers to software-based synthesizers or virtual instruments. Now, there is a large and growing number of software-based instruments that are compatible with the Pro Tools environment.

Native or DSP

Each plug-in is composed of complete computer code that forms a sound-processing module or a synthesis-sound module. Some are processed by the host computer's main processor. This system is called native processing. Others are written to be processed by special *digital signal processor* (DSP) chips that reside on additional cards that have been installed in the computer. This is the case with *time division multiplexing* (TDM) and TDM II (HD) Pro Tools cards.

Pro Tools supports several formats of audio plug-ins.

CHAPTER 4

Dynamics and Levels

Dynamics Processors

Dynamics processors, such as compressors, limiters, expanders, and noise gates, are some of the most important tools for engineers. They can also be difficult to understand, due to the wide range of applications and the sometimes subtle, sometimes extreme effects they have on the mix.

Compression is the most-used dynamic processing technique in most music today, and many would argue that it is the biggest single factor in the overall sound of a mix. There are different schools of thought on how much compression is enough and how much is too much. But generally, the compressed sound has become more and more popular over time.

Dynamics sometimes gets confused with *loudness*. Technically, *dynamics* refers to the difference in amplitude between the loudest and softest sounds in a track. When sounds are referred to as dynamic, it means there is a lot of variability in the amplitude.

Compression is used to make tracks less dynamic by leaving the quiet parts untouched and by reducing the amplitude of the loud parts. This actually reduces the overall amplitude of the track. So, after the signal is compressed, its overall volume is usually boosted. A typical application of compression is to even out the volume of a rock bass guitar track and then boost it in so that it provides a more consistent level to anchor a mix. Another common use is to apply some compression to the final mix in pop music to make the volume levels more consistent and powerful for radio play.

Typically, compressors have six controls, although some popular hardware compressors do not have all of them, as some fixed settings are what give certain instruments or songs their characteristic classic sounds:

- *Input Gain* controls the signal level going into the compressor.

- *Threshold* controls the level at which the compressor starts reducing the volume.

- *Ratio* controls how much the volume will be reduced.

- *Attack* controls the length of time that the compressor waits before reducing the volume.

- *Release* controls the length of time before the compressor allows the volume to return to normal.

- *Output* or *Makeup Gain* controls how much volume boost is applied to the signal after it is compressed.

These six effects can offer a fantastic amount of modification over the sound.

Limiters, gates, expanders, multiband compressors, de-essers, normalizers, and ultramaximizers are all variants of the basic compressor and can be used to improve a mix.

The limiter is a kind of compressor designed to provide control over the hottest peaks in a signal. Unlike a compressor, the goal of the limiter is to change the dynamics as little as possible. Used properly, the limiter will cut off the loudest peaks in the signal without changing the dynamics of the rest of the audio. Hardware limiters typically have only two controls—input and threshold. There is also a release control on some limiters. If the ratio control on a compressor is set to 10:1 or higher, it is essentially performing as a limiter.

A gate provides the opposite function. Rather than attenuating loud signals, the gate attenuates only the soft signals. It's often called a noise gate because it is usually used to eliminate noise when the desired signal is not present. A gate can be used to eliminate microphone bleed by setting it so that low-level, off-mic signals do not pass. It can also be used to eliminate the hum from a guitar amp by setting the threshold above the level of the hum.

Multiband compressors compress audio within certain frequencies. Where an equalizer attenuates all instances of a specified frequency, a multiband compressor attenuates the specified frequency only when the threshold is crossed.

A de-esser is a band-specific compressor that is designed specifically for the sibilant frequencies produced by human speech. De-essers react to frequencies in the range of 2 to 16 kHz. The range around 3 to 6 kHz is the most important because the ear is most sensitive to those frequencies, and they can sound really nasty when over-amplified. Because it is a compressor—not an *equalizer* (EQ)—the de-esser only reacts to excessive treble. The de-esser can really help your vocals. When used with a good EQ, you can boost the overall treble to add breath and sparkle while controlling excessive treble only on the consonants that become shrill. The overall effect is a wetter, shinier sound that never gets harsh.

Expanders can work in a couple of different ways. A downward expander attenuates the signal as it passes below the threshold, similar to the way a gate cuts off a signal that passes below the threshold. An upward expander is the reverse of a compressor. It amplifies the peaks of a signal that pass its threshold. It enhances the dynamics of the loudest signals and can be used to add punch or attack in a muddy track.

Multiband Dynamics Tool (MDT)

Antares Audio Technologies
www.antarestech.com

The MDT is a *time division multiplexing* (TDM) *digital signal processor* (DSP) plug-in that has a wide range of applications (see Figure 4-1). It can function as a multi-band compressor, limiter, expander, or de-esser. It is highly programmable, with up to 30 separate thresholds and ratios.

The plug-in can work as a full band processor or in 3-band or 5-band multiband modes. The multiband filter-mode menu controls the number of bands into which the signal is divided before processing. Multiband modes divide the frequencies into equally spaced bands.

The Input Offset feature allows the relationship of each band to the I/O curve to be adjusted individually. Each band can be processed on a separate portion of the I/O curve.

Figure 4-1
The MDT window

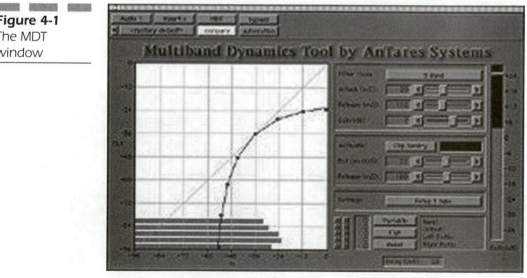

This plug-in has applications in mastering, post production, broadcast, track sweetening, sound effects, and sample editing.

Drawmer Dynamics

Drawmer
www.digidesign.com

The Drawmer Dynamics TDM plug-in is based on the Drawmer DS201 Noise Gate and Drawmer DL241 Compressor/Limiter hardware units (see Figure 4-2). It emulates these devices closely and has many automatic functions, just as the hardware units do. Drawmer's hardware units pioneered several concepts that have now appeared in the software version. These include frequency-conscious noise gating that uses variable high-pass and low-pass filters and comprehensive envelope control. It also has the ability to trigger the noise gate from a key signal using another audio track within Pro Tools as the controlling source.

Figure 4-2
Digidesign's
Drawmer
Dynamics

Bomb Factory Classic Compressors

Bomb Factory
www.bombfactory.com/

Bomb Factory Classic Compressors are realistic emulations of the LA-2A and 1176 vintage compressors used in top pro studios (see Figure 4-3). These are the original software models of these units and the modeling technology developed by Bomb Factory, along with their thorough understanding of the original hardware and its applications, add up to some very accurate and effective tools.

Bomb Factory LA-3A

Bomb Factory
www.bombfactory.com/

The LA-3A leveling amplifier is the latest addition to the Classic Compressors line (see Figure 4-4).

Figure 4-3
The Bomb
Factory Peak
Limiter

Figure 4-4
The Bomb
Factory LA-3A

The LA-3A is known for adding a smooth texture to many different instruments and voices.

Fairchild Model 660

Bomb Factory
www.bombfactory.com/

The hardware Fairchild 660 is a variable-mu tube limiter that has been in use since the 1950s. It was a very expensive unit that used tubes for gain reduction. The Bomb Factory plug-in is designed to emulate the hardware as closely as possible (see Figure 4-5).

Bomb Factory JOEMEEK SC2 Compressor

Bomb Factory
www.bombfactory.com/

The JOEMEEK SC2 is a powerful compressor that can be used to create more extreme compression effects (see Figure 4-6). While some compressors attempt to retain a sonic transparency, this one is designed to add a musical effect.

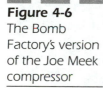

Figure 4-5
The Fairchild
660 emulation

Figure 4-6
The Bomb
Factory's version
of the Joe Meek
compressor

Maxim

Digidesign
www.digidesign.com

Maxim is a high-performance limiter that is primarily
designed for stereo mastering (see Figure 4-7). It optimizes
the level of the tracks while effectively limiting any peaks
or spikes. It also performs a dithering function. In addition
to use in mastering, Maxim can also serve as a dynamics
processor.

Figure 4-7
The Maxim
window

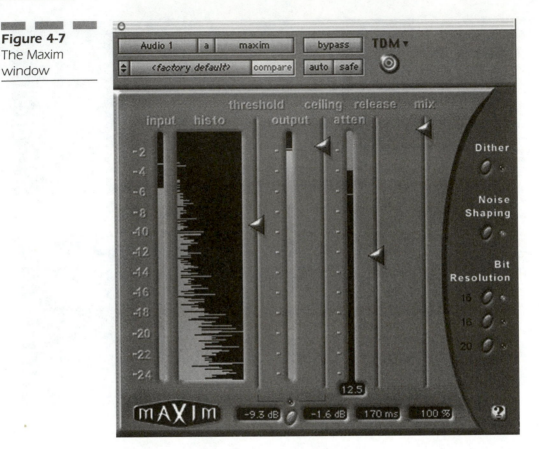

Omnipressor

Eventide
www.eventide.com

Eventide Clockworks created many of the hardware
devices that defined the sounds of classic rock. The Omni-
pressor was a serious compressor that became very popular in the 1970s and 1980s.

This software version is an emulation of that classic
sound and should be used as an effects compressor, espe-

Figure 4-8
The digital version of Eventide Clockwork's hardware

cially in high-volume guitar tracks and other high-volume electric tracks (see Figure 4-8).

Compressor Bank

McDSP
www.mcdsp.com

Compressor Bank is a set of eight plug-ins. CB1 is the basic compression unit, CB2 is for filtered compression, CB3 is modeled compression, and CB4 is a set of preset compressors and limiters. They come in both stereo and mono configurations.

Standard controls include Output (make-up gain), Threshold, Compression (Ratio), Attack, and Release. In addition to the standard Threshold and Compression (Ratio) controls, the actual shape and response of the compression curve can be adjusted with the Knee and Bite controls. Knee softens the compression curve, creating a smoother response. Bite gives the compressor the ability to allow signal transients to pass uncompressed, while the overall compression response is unchanged (see Figure 4-9).

The concept is that these controls enable the user to emulate the characteristics of various vintage compressors.

Figure 4-9
The Compressor
Bank window

MC2000

McDSP
www.mcdsp.com

The MC2000 is like a multiband version of the CB1 module of Compressor Bank described above. It is a set of six plug-ins, configured in 2, 3, and 4 bands in both stereo and mono versions (see Figure 4-10).

PSP Vintage Warmer

PSP Audioware
www.pspaudioware.com

The PSP Vintage Warmer plug-in is a digital simulation of an analog single- or multiband compressor/limiter (see Figure 4-11). As the name implies, it imparts a warmth associated with tube compressors and analog tape

Figure 4-10
The MC2000

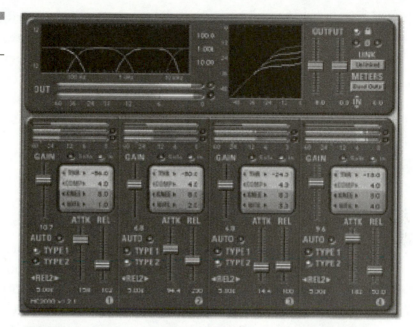

Figure 4-11
The PSP Vintage
Warmer

machines. It can perform single- and multiband compression, as well as brick-wall limiting. It is known for its overload characteristics, which allow the processor to generate saturation effects typical of analog tape recorders. It can be used as a mastering tool or for effects in tracks.

Compressor X

Sonic Timeworks
www.sonictimeworks.com

Compressor X is an RTAS plug-in that is designed to emulate a wide range of classic and modern compressors. It is highly automatable and supports both compression and brick-wall limiting. It has very fast attack times and a responsive analog type meter (see Figure 4-12).

Sony Oxford OXF-R3 Dynamics

Sony
www.sonyplugins.com

The Sony Oxford OXF-R3 takes the channel dynamics capabilities of the famous Oxford console and delivers them in a software plug-in format.

Figure 4-12
The Compressor
X window

There are separately controlled sections for the compressor, limiter, gate, and expander (see Figure 4-13). There are very wide control ranges offered within each of the applications. The *graphical user interface* (GUI) is sophisticated enough to enable creative applications without loss of control. It also has excellent sonic characteristics.

This plug-in has a gain-controlling side-chain element, which works by evaluating the program level at the input and calculating the required output gain by dead reckoning. Also, look-ahead processing allows gain control to be initiated in advance of the signal without signal quality loss.

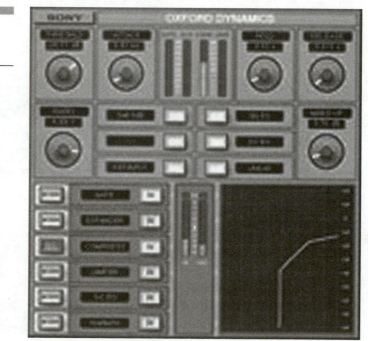

Figure 4-13
The Oxford
plug-in

1176LN

Universal Audio
www.uaudio.com

This plug-in is Universal Audio's rendition of the classic 1960s 1176LN compressor (see Figure 4-14). It is designed for Pro Tools TDM HD and mix systems. It is fully automatable, supports stereo and mono operation, and has a full set of manual controls. The 1176LN can run one instance at 44.1 kHz and 48 kHz on both the mix and HD systems. It will not run at 96 kHz.

Included with the 1176LN is the 1176SE compressor (see Figure 4-15). This compressor is a less DSP-intensive version of the 1176LN. The 1176SE can run six instances on HD at 44.1 kHz and 48 kHz, and four instances in the mix system. On HD systems, it will run three instances at 96 kHz or one at 192 kHz.

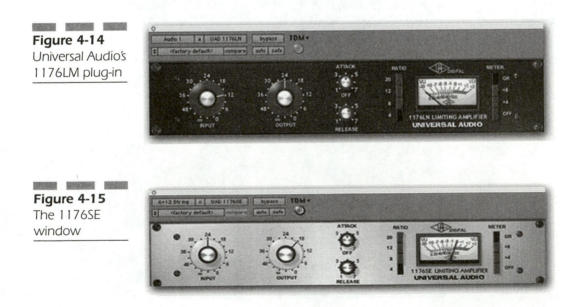

Figure 4-14
Universal Audio's
1176LM plug-in

Figure 4-15
The 1176SE
window

LA-2A

Universal Audio
www.uaudio.com

This is Universal Audio's version of the Teletronix LA-2A leveling amplifier. The interface looks like the analog original; it supports an emulation of the optical attenuator that defined the original machine (see Figure 4-16). It has a full set of controls and full automation capability.

Renaissance Channel

Waves
www.waves.com

Renaissance Channel is a precision channel processor module of Wave's Renaissance collection of plug-ins. It has equalization, compression, limiting, and gating capabilities.

The plug-in has vintage-modeled EQ and compressor emulations, selectable EQ/Dynamics ordering, and overload protection (see Figure 4-17).

Figure 4-16
The LA-2A is designed to look like the original.

Figure 4-17
The Renaissance
Channel
module

Renaissance Compressor

Waves
www.waves.com

Renaissance Compressor is a part of Waves Renaissance
Collection. It is a simple high-performance program for
vocal or instrument processing. It incorporates a brick-

wall internal limiter. A unique warm character setting adds low-frequency harmonics while approaching deeper compression to produce a warm, rich sound. Numerous factory presets emulate the performance of various popular compressors. However, the interface is not an emulation, but is a simple, intuitive control set that is designed to be more musical than technical (see Figure 4-18).

Vox

Waves Renaissance
www.waves.com

Renaissance Vox is a compressor/limiter based on a hardware unit that originally had only one knob. This one

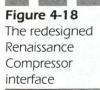

Figure 4-18
The redesigned Renaissance Compressor interface

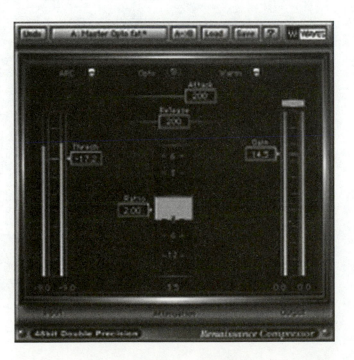

actually has two controls, due the addition of a gate, and is ideal for vocal tracks (see Figure 4-19).

The compressor, gate, and precision limiter have automatic gain staging for clip-free maximization. The energy control adjusts the threshold of the multistage downward expander, and the compression control adjusts the amount of compression and limiting while automatically providing output gain boost. The one-control format for compression, limiting, and level maximization make it simple and effective.

Figure 4-19
The Vox UI

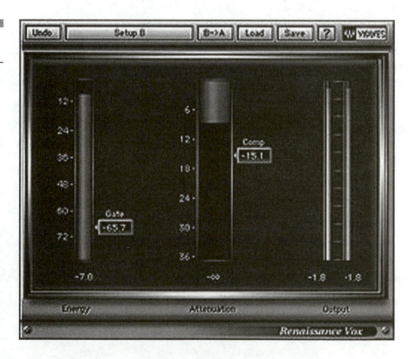

EQ/EQ Dynamics

Aphex Aural Exciter and Aphex Big Bottom Pro

www.digidesign.com

The hardware version of the Aphex Aural Exciter is the industry standard. This TDM plug-in version brings the hardware concept to Pro Tools. The original hardware was introduced in 1975. There have been improvements introduced in subsequent models. The current hardware, with some improvements over the original, is the model for the plug-in (see Figure 4-20).

Figure 4-20
Aphex plug-ins imitate their hardware couterparts.

The Aural Exciter increases the presence of instruments and vocals up out of the mix by increasing bandwidth without increasing level. The resulting improved presence, clarity, speech intelligibility, and detail do not significantly increase peak output.

Big Bottom is modeled after the Big Bottom circuit in the Aphex Model 104. It enhances bass frequencies without adding peak level. It improves bass resonance and bass articulation, and adds extended low frequencies.

JOEMEEK VC5 Meequalizer

Bomb Factory
www.bombfactory.com/

The Bomb Factory JOE MEEK VC5 Meequalizer is a unique treble-and-bass circuit with a sweepable mid control. The plug-in is based on an original Joe Meek hardware piece (see Figure 4-21).

Pultec EQP-1A

Bomb Factory
www.bombfactory.com/

The Pultec EQP-1A is an emulation of the rare and classic hardware with the same name. The original was a true

Figure 4-21
The VC5
Meequalizer

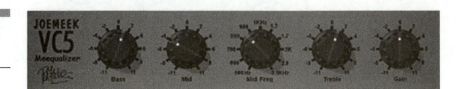

Figure 4-22
The EQPIA is
easy to control.

tube EQ that was a staple of high-end recording and mastering studios. It was not a surgical EQ, but was designed to sweeten high and low frequencies and generally make everything that passed through it sound better. It can be used on individual tracks or across an entire mix.

This plug-in is very easy to control, with separate boost and attenuation controls (see Figure 4-22). The Pultec has long been a choice of recording and mastering engineers for its ability to bring out individual frequency ranges without significantly altering other frequencies.

Focusrite d2/d3

Focusrite
www.digidesign.com

Focusrite d2 is a 24-bit EQ modeled after the Red Range 2 Dual EQ hardware unit (see Figure 4-23). It has three different stereo or mono modules (six band, four band, and dual or single mono band) and includes high- and low-pass filters, high- and low-shelving filters, and high-mid and low-mid peaking filters.

Focusrite d3 is a compressor/limiter modeled after the Red 3 hardware unit. It provides two mono or stereo configurations, as well as flexibility and versatility through separate pre- and post-fader insert configurations. A floating Editor window provides access to both compression and limiting functions. There is excellent feedback from the interface with numeric displays located below each rotary knob and the plasma-type meters (see Figure 4-23).

MDW Hi-Res Parametric EQ

Massenburg Design Works
www.digidesign.com

George Massenburg is an industry legend and this program is his high-end EQ plug-in solution for Pro Tools (see Figure 4-24).

The high resolution of this plug-in produces sonic clarity. It is based on 48-bit processing and high-resolution 96 kHz sampling rate processing. With the unit's double 48-bit processing, the audio is processed at twice the sampling rate with optimal precision, resulting in less noise, increased headroom, fewer artifacts, and wider dynamic range. Even when working in sessions set to 44.1 or 48 kHz, the MDW EQ processes the audio at 88.2 and 96 kHz, respectively. The result is high-resolution audio, fewer artifacts, and more predictable filter curves.

The plug-in emulates the GML 8200's constant-shape, reciprocal-filter curves. The GML 8200 is a hardware five-band parametric equalizer that is considered to be the industry standard reference for filter curves. The unit has a wide frequency selection from 10 Hz to 41 kHz.

Figure 4-24
The MDW Hi-Res
Parametric EQ

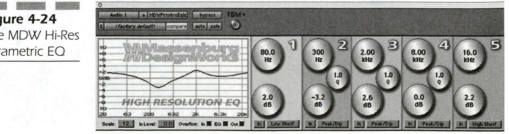

FilterBank

McDSP
www.mcdsp.com

FilterBank is a set of 20 EQ/filter plug-ins consisting of
2-, 4-, and 6-band configurations of equalizers and filters
in mono and stereo versions (see Figure 4-25).

The included equalizer and filter design parameters
emulate vintage and modern EQs, both solid-state and
tube. They can be used as is or can be edited to create custom EQs.

The adjustable control and design parameters include
high- and low-shelving EQ with independent control of
frequency, gain, peak, slope, dip, parametric EQ, low- and
high-pass filters, band pass and band reject filters,

Figure 4-25
McDSP's Filter-
Bank

a frequency control range from 20 to 21,000 Hz, and analog saturation modeling in each EQ/filter band to prevent digital clipping.

Equalizer

Serato Audio Research Ltd.
www.serato.com

The Serato Equalizer is a new multiband graphic equalizer, with a real-time spectrum analyzer and advanced curve editor that is not yet released (see Figure 4-26). It will initially be released as an RTAS and HTDM plug-in for Pro Tools.

Figure 4-26
Serato Audio's Equalizer features a spectrum analyzer.

Channel X

Sonic Timeworks
www.sonictimeworks.com

Channel X is a new RTAS channel strip with parametric EQ and built-in dynamics. It includes low- and high-pass filters, four bands of sweepable EQ, and an active-compression dynamics section with threshold, ratio, and release controls. All parameters are automatable and each EQ band is individually bypassable. The plug-in has LED-style meters for input and gain reduction (see Figure 4-27).

L2-Ultramaximizer

Waves
www.waves.com

The Waves L2-Ultramaximizer software is a software emulation of the L2 hardware processor (see Figure 4-28).

L2 can be used to significantly increase the average signal level of typical audio signals without introducing audible side effects. It has a wide range of applications, including mastering, mixing, and recording. It can also be used for extreme limiting, intentional pumping, and vintage dynamic-processing effects.

Figure 4-27
The Channel X window

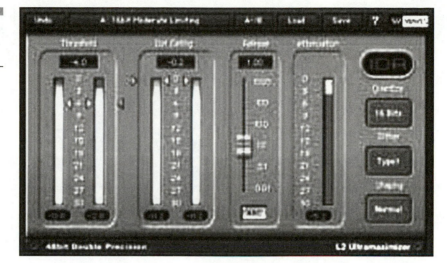

Figure 4-28
The L2
Ultramaximizer

The plug-in's core technology features include a brick-wall look-ahead peak-limiter algorithm, a proprietary dithering technology with ninth-order noise shaping, and auto-release control.

Maxx Bass

Waves
www.waves.com

MaxxBass extends the perceived bass response by adding a series of bass harmonics to the signal that stimulate a psychoacoustic bass-enhancing effect. This technique maximizes a known acoustical phenomenon employed in small speaker design to make them capable of producing audible bass.

The bass frequencies are enhanced without traditional EQ or bass compression. With Maxx Bass you can actually

Figure 4-29
The Maxx Bass
window

Figure 4-29
The Maxx Bass window

remove the original bass signal completely and the added harmonics will effectively recreate it to the ear.

This plug-in can be used to sharpen muddy bass lines without disturbing the other instruments. It is also useful for mastering or site-specific mixing (see Figure 4-29).

Q10 ParaGraphic EQ

Waves
www.waves.com

The Q10 ParaGraphic EQ is a powerful equalizer, with 200+ audio effects and processes in an included setup library with everything from extreme notch filters to precision mastering, including pseudo-stereo, band limiting, harmonic comb filter, pre/de-emphasis, crossover, notch, multimedia, and other EQs (see Figure 4-30). The plug-in has up to 10 bands of true parametric or high/low shelving EQ, as well as high-pass (low-cut) or low-pass (high-cut) filters. It can be used as two independent mono equalizers or one stereo EQ.

With the Q10 ParaGraphic EQ, subtle to extreme equalization is possible. The controls include on/off, filter type, gain, frequency and Q controls for each band, left/right/strap modes, and left/right input/output faders. The stereo-input faders have switches for reversing channel polarity. The Q10 has an uncolored, clean, predictable sound.

Figure 4-30
The ParaGraphic EQ

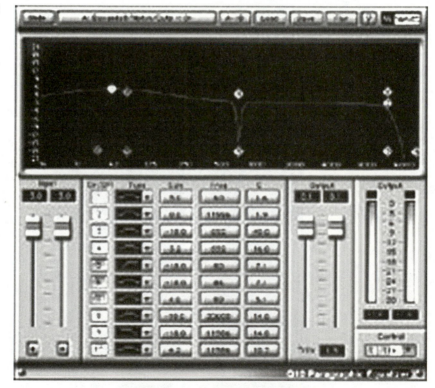

Renaissance Bass

Waves
www.waves.com

Renaissance Bass is the second-generation implementation of Maxx Bass technology (see earlier entry). It uses the same psychoacoustic concepts. The user can selectively add harmonics that significantly enhance the bass perception by the listener. Renaissance Bass provides greater effectiveness and simpler operation than Maxx Bass. It controls clips better and has an easier-to-use interface (see Figure 4-31).

Figure 4-31
The easy-to-use Renaissance Bass UI

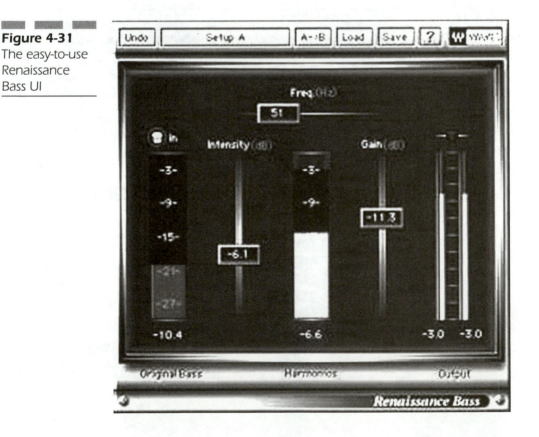

Renaissance DeEsser

Waves
www.waves.com

The Renaissance DeEsser uses technologies from C4, Rvox, and Waves DeEsser, and then adds some of its own features. It uses the same phase-compensated crossover as the C1 and C4 compressors to decrease coloration and phase-modulation compressor artifacts. It has an adaptive threshold feature and it dynamically adjusts to the input signal to provide a smoother response. The user can control the maximum gain reduction by setting the attenuation range control. The graphical display provides excellent user feedback (see Figure 4-32).

Figure 4-32
The Renaissance
DeEsser
window

Renaissance Equalizer

Waves

www.waves.com

Renaissance Equalizer is a double-precision resolution equalizer. It is designed to emulate analog-style EQs. Its simple control features include real-time graphing, preview, and bypass functions (see Figure 4-33).

Figure 4-33
The Renaissance
Equalizer

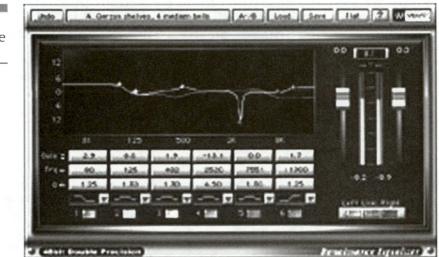

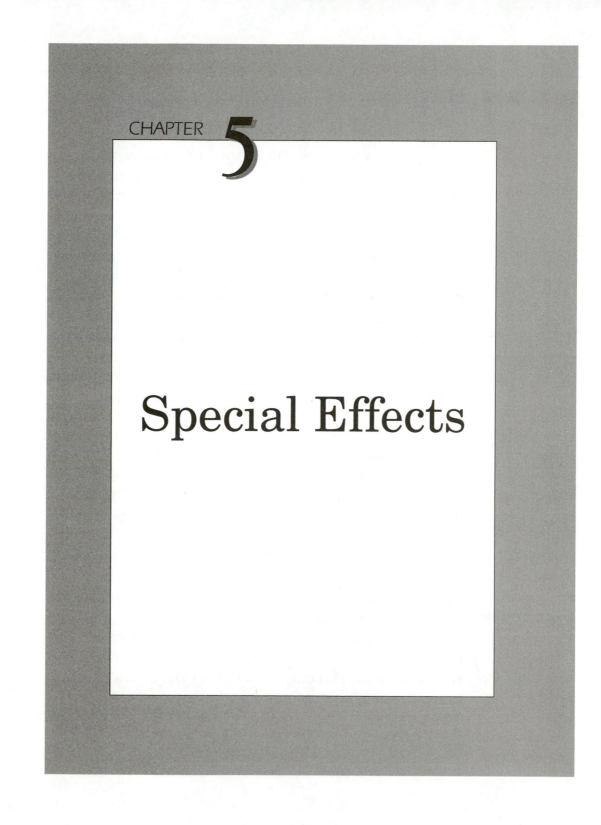

CHAPTER **5**

Special Effects

Reverb

Reverb is the short name for *reverberation* and is one of the most heavily used effects in music. It is generally thought of as the result of the many reflections of a sound that occur in a room. There is a direct path from a sound source to our ears but, in most cases, sound waves also take a slightly longer path, reflecting off of walls, floors, and ceilings before arriving at the ear. The reflected sound waves arrive a little later than the direct sound, because it travels a longer distance, and the delayed sound generally has different characteristics than the direct sound. It is a little weaker because it has partially been absorbed by the surfaces in the room. The primary reflections are sometimes called early reflections. Typically, in a real room, reflected waves bounce off multiple surfaces before arriving at the ear. You hear not only the direct sound, but also a series of delayed and attenuated sound waves, sometimes referred to as late reflections or diffuse reflections. The delayed and attenuated sound waves, or reverberation, of the original sound are defined by the dimensions and materials of the room or space in which the sounds are generated and heard.

Reverb is a little different than an echo effect. An echo is usually a distinct, delayed version of a sound—the delays are long and can be perceived as separate sounds. The echoes are far enough apart that you do not usually get the room effect that you get with reverb. With reverb, the delay is very short, so that it is not usually perceived as a separate sound and only the effect of the series of reflections is heard.

There are many different kinds of reverbs and various types of reverb devices in the physical, analog world. The available software-based reverbs for Pro Tools are also quite diverse.

D-Verb

Digidesign
www.digidesign.com

D-Verb is a basic workhorse reverb that comes with all Pro Tools systems (see Figure 5-1). It comes in *time division multiplexing* (TDM), RTAS and AS formats, as well as mono and stereo versions. D-Verb is easy to use and produces high-quality results. For more info, check the web site at www.digidesign.com/.

Figure 5-1
D-Verb

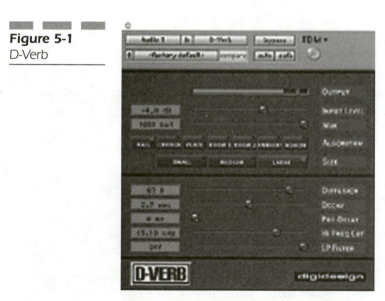

Reverb One

Digidesign
www.digidesign.com

Reverb One is Digidesign's highest-quality reverb plug-in
(see Figure 5-2). It is available for TDM, TDMII, and AS.
For more info, check the web site at www.digidesign.com/.

Altiverb

Audio Ease
www.audioease.com

Altiverb is a sampled room reverb. It relies on the Mac G4
Altivec engine for processing. Altiverb ships with samples

Figure 5-2
Digidesigns
Reverb One

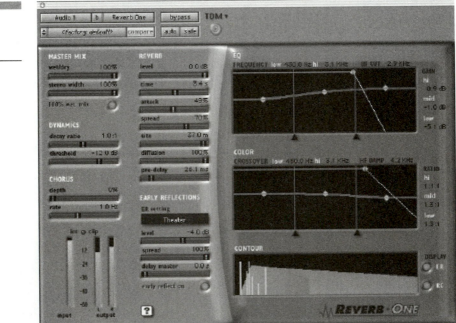

of dozens of acoustical environments. These include concert halls, recording studios, castles, and many other spaces with unique acoustic profiles from around the world.

An Altiverb preset called an *impulse response* (IR) creates reverb by generating sounds in an acoustic space and then recording a digital audio file of the resulting reverberation decay that is characteristic of that acoustic space. The original, dry source signal is then extracted from the audio file, leaving behind the acoustic fingerprint of the space. The IR file is then placed in a folder where it appears as a preset inside the Altiverb plug-in. It can be applied to any audio signal as real-time reverb. Besides the IRs that ship with Altiverb, many additional samples from a growing collection are downloadable from the web site at www.audioease.com/IR.

Altiverb also includes the capability to record your own sounds and create your own unique IRs. You can also add a picture of the sampled space that will appear in the Altiverb interface by taking a digital photo and putting with the IR as a JPEG file (see Figure 5-3).

Figure 5-3
The Altiverb interface with a digital photo

Reverb X RTAS

Sonic Timeworks
www.sonictimeworks.com

Reverb X RTAS is a Mac/PC Pro Tools-compatible version
of a plug-in that was formally only available for PCs. It's
HD compatible at up to 192 kHz and features an
analog graphical meter. It's a simple, clean interface (see
Figure 5-4).

DreamVerb

Universal Audio
www.uaudio.com

DreamVerb uses a graphical interface to create room models. There are adjustable settings for room materials and a
graphic menu of room shapes and sizes that can be manipulated in real time (see Figure 5-5).

There are extensive parameter controls for the intensity,
timing, and onset of early reflections and late-field reverberation. There are also controls for the density of the
air, which can be changed to simulate different ambient
situations.

Figure 5-4
The Reverb X
RTAS interface

Figure 5-5
DreamVerb by
Universal Audio

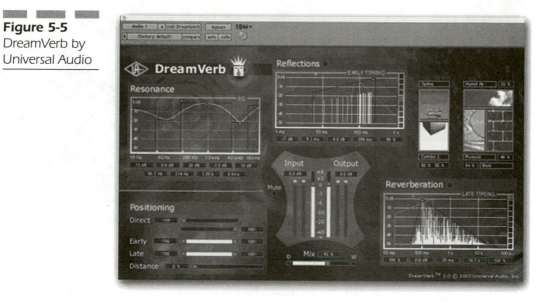

Figure 5-5
DreamVerb by
Universal Audio

The plug-in includes a large list of different materials and room shapes, which can be customized further by blending or morphing the different room shapes and surfaces.

DreamVerb also features a five-band active EQ. All parameters can be adjusted in real time.

Renaissance Reverberator

Waves

www.waves.com

The Renaissance Reverberator is designed to be simple and provide classic reverb sounds (see Figure 5-6). It has a characteristic sound that is dense and rich in texture, with an emphasis on second-generation early reflections. It is very functional when classic reverb sounds are essential.

Figure 5-6
The Renaissance
Reverberator

The Renaissance Reverberator comes in several of Waves' large plug-in collections and is available for all formats. The Renaissance Collection also includes a compressor and equalizer.

TrueVerb

Waves
www.waves.com

TrueVerb, a TDM/TDM-HD plug-in available in bundles from Waves, is a room-modeling type of reverb that is designed for multichannel mixing. It maintains the level and balance of the mix and does not disturb the stereo image. User controls include room size, decay times, frequency response, and sound-source placement. There are separate controls for early reflections and reverb, and the effect is fairly transparent in that it provides a convincing reverb effect with little additional coloration (see Figure 5-7).

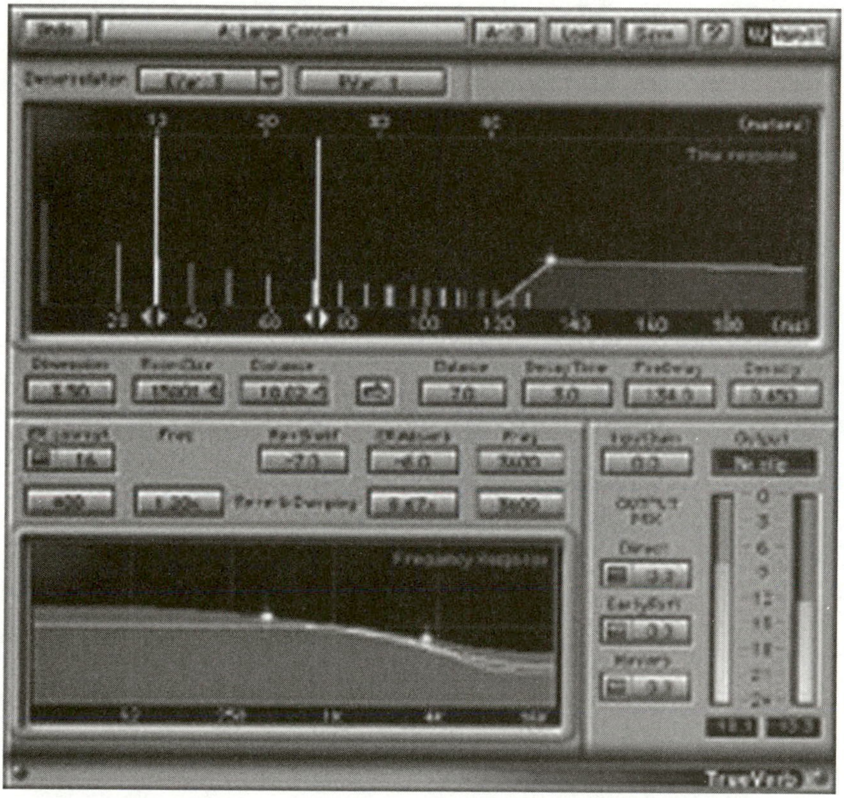

Figure 5-7
The TrueVerb
room-modeling
plug-in

There is a comprehensive setup library that is driven by embedded psychoacoustic rules that tend toward natural sounds. Extensive additional controls can be used to create more creative effects. The Sound Distance control is somewhat unique.

Effects Processors

Effects processors come in a variety of forms. Many are emulations of earlier hardware devices and allow you to add a distinct character to your projects.

Moogerfooger 12-Stage Phaser

Bomb Factory
bombfactory.com

The moogerfooger 12-Stage Phaser combines a switchable 6- or 12-stage phaser with a wide-ranging variable *low-frequency oscillator* (LFO) (see Figure 5-8).

Moogerfooger Analog Delay

Bomb Factory
www.bombfactory.com

Bomb Factory's moogerfooger Analog Delay is a plug-in that emulates the warm-sounding classic moog analog delay (see Figure 5-9).

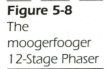

Figure 5-8
The
moogerfooger
12-Stage Phaser

Figure 5-9
The analog delay plug-in emulates the original.

Moogerfooger Lowpass Filter

Bomb Factory moogerfooger
www.bombfactory.com

The moogerfooger Lowpass Filter features a 2-pole/4-pole variable-resonance filter with envelope follower. It adds warm, fat analog resonance on guitars, basses, or other instruments (see Figure 5-10).

Ring Modulator

Bomb Factory moogerfooger
www.bombfactory.com

The moogerfooger Ring Modulator has a wide-range carrier oscillator and a dual sine/square waveform LFO (see Figure 5-11).

SansAmp PSA-1

Bomb Factory
www.bombfactory.com

Bomb Factory's SansAmpTM PSA-1 emulates the Tech 21 outboard hardware with the same name. The hardware was originally designed as a preamp that had the flexibility to achieve a wide range of miced tube amp-speaker cabinet sounds. The software comes with the original 49 presets with sounds from classic amps from ampmakers such as Marshall, Mesa Boogie, Hiwatt, Fender, and Ampeg, plus many more new presets (see Figure 5-12).

Figure 5-12
Bomb Factory's version of SansAmp

Tel-Ray

Bomb Factory
www.bombfactory.com

In the 1960s, a small California company created a delay device that involved a can filled with electrolytic oil and a tiny pickup on a flywheel that sloshed around in the oil, producing a unique echo effect. It eventually evolved into what was known as the Adineko design and found its way into effects devices from numerous manufacturers. Bomb Factory's Tel-Ray emulates that classic technology digitally (see Figure 5-13).

Voce Chorus/Vibrato

Bomb Factory
www.bombfactory.com

Voce Chorus/Vibrato is a recreation of the Hammond B3 organ's mechanical scanner vibrato with three settings each for vibrato and chorus (see Figure 5-14).

Voce Spin and Chorus/Vibrato

Bomb Factory
www.bombfactory.com

This plug-in is a simulation of the famous Leslie rotating speaker. It's a great chorus effect with 15 presets (see Figure 5-15).

Figure 5-13
The Tel-Ray
interface

Figure 5-14
The Voce
Chorus/Vibrato

Figure 5-15
Voce Spin and
Chorus/Vibrato

H910

Eventide
www.eventide.com

The H910 is a plug-in version of the original H910 Harmonizer, which was discontinued in 1984. It features pitch change, reverb, delay, and a variety of special effects (see Figure 5-16).

The plug-in can move pitch an octave up or down while preserving harmonic ratios and create additional notes at perfect intervals, allowing a range of harmonizing effects and vocal multiplying. The H910 can be used for doubling vocals, as well as several types of reverb and echo.

The simultaneous use of feedback, delay, and pitch change can be used to create a variety of audio effects. Maximum delay and one interval of pitch change combined with feedback generate a musical progression of a single note. Pitch change and feedback with no delay give a robotic speech effect. A short delay with feedback but no pitch change gives a hollow flanging or tunneling effect, and a long delay generates a distinctive reverb-type effect.

Figure 5-16
The H910
plug-in

Instant Flanger

Eventide
www.eventide.com

This plug-in replicates the Instant Flanger, an analog flanger, first released in 1976 and sold through 1984 (see Figure 5-17). Flanging is an analog effect created when two tape machines are playing identical material and the flange of one of the open reels is touched. The resulting sound is a distinct roar that has been used in many rock recordings. All kinds of hardware flangers are used today, especially for guitar.

Instant Phaser

Eventide
www.eventide.com

The Instant Phaser plug-in is modeled after the world's first analog phaser that was introduced in 1971. It was used heavily in classic rock recording of the 1970s. It is a single-function processor with a sweeping filter bank with two outputs that are 180 degrees out of phase from each other. It supports 48, 96, and 192 kHz sample rates.

Figure 5-17
The Instant
Flanger
interface

MindFX Volume 1

GFXpansion
www.fxpansion.com

MindFX Volume 1 is a collection of six sound-design effects — Autopole 2, Mtap 2, MIDIcomb 2, Phatsync 2, Ringmod 2, and RobotikVocoder 2.

Autopole 2 is a stereo dual-filter bank, with two variable-waveform LFOs, two independent envelope generators, signal trackers, flexible modulation and signal routing, vector mod controls, and a variety of multimode filters (see Figure 5-18).

Mtap 2 is a multitap delay plug-in with x/y editing of time, frequency, amplitude, and pan. It has two onboard LFOs, saturation/degrade processors, and feedback routing for complex delay effects.

Figure 5-18
The Mind FX
Volume 1

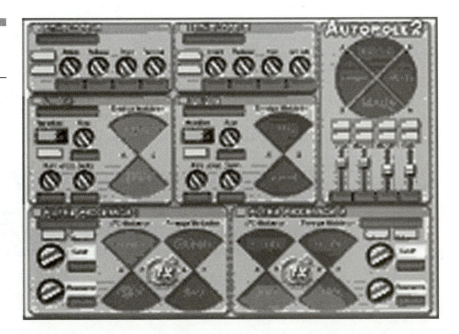

MIDIcomb 2 is a polyphonic comb filter that can be driven by live or prerecorded audio.

Phatsync 2 is a pattern-controlled filter bank. Phatsync2 has two onboard pattern sequencers and is ideal for groove-syncing.

Ringmod 2 has four independent synths as mod sources, multiband operation, multiple performance modes, envelopes, LFOs, and performance controls. It can create ring modulation, frequency shift, and spectral dissonance effects playable from a *musical instrument digital interface* (MIDI).

RobotikVocoder 2 is a 24-band vocoder. It can perform crossfades and morphs between different filter-bank setups. It has onboard dynamics and delay. It is MIDI-controllable with an onboard two-oscillator polyphonic synthesizer.

MindFX Volume 2

GFXpansion
www.fxpansion.com

MindFX Volume 2 is another collection of six sound-design effects, including Snippet Resynth, Streamsampler, FuZzX, Hyper Channel, Helicopter, and Evolvor (see Figure 5-19).

Snippet Resynth is a real-time resythesizer that performs short-cycle real-time granular resynthesis with wave-table sequencing. It has four independent, polyphonic grain operators with independent input and output frequency ranges; each grain operator also has its own amplitude envelope.

The sum of the grain operators is fed through a multimode filter and a delay unit.

Figure 5-19
Mind FX
Volume 2

Streamsampler is an interactive, four-channel, audio capture, processing, and playback engine that is entirely controlled in real time from a MIDI keyboard or sequencer.

FuZzX is a multiband distortion, degrade, saturation, and noise processor.

Hyper Channel includes four fairly standard channel processes: gate, compressor, EQ, and reverb. Each processor has 12 internal states, and each state is assigned to a MIDI key.

Helicopter is a multiband semimodular effects processor that incorporates tremolo, vibrato, chorus, phase, flange, autopan, mod-delay, pan-delay, stereo widen, and other effects.

Evolvor is a "morphing variable-O time-domain convolution processor." It is a new kind of filter that cuts and boosts frequencies in unique ways.

Quad Frohmage Bundle

Ohmforce
www.ohmforce.com

The Quad Frohmage Bundle from Ohmforce is shown in Figure 5-20.

AmpliTube

IK Multimedia
www.amplitube.com

AmpliTube is a guitar amp and effects modeling plug-in. It has separate preamp, EQ, amp, cabinet and mic

Figure 5-20
The Quad
Frohmage
plug-in

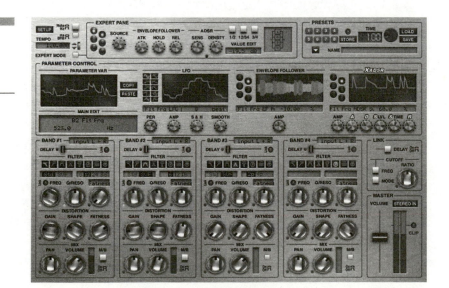

Figure 5-21
The AmpliTube
guitar effects
plug-in

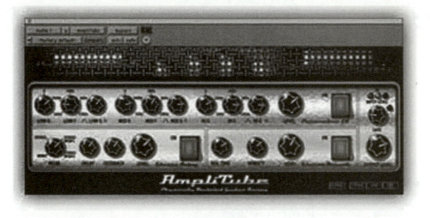

modeling with 1,260 different amp configurations that can be emulated. It also models various stomp boxes and post effects and has unique amp-blending capabilities (see Figure 5-21).

Amp Farm

Line 6
www.digidesign.com

The Amp Farm TDM plug-in can process live guitar inputs or prerecorded tracks right up to the final mix (see Figure 5-22). You can record a clean guitar while hearing the amp effect in the headphones, and then either use the original amp setting in the mix or apply a new one.

The plug-in features amp models based on Fender Twin, Blackface Deluxe Reverb and Bassman, Marshall JTM 45, Plexi and JCM 800, Vox AC 30, Mesa Boogie Dual Rectifiers, Soldano SLO and X88R, and Matchless Chieftain.

The maximum supported sample rate for the Amp Farm plug-in is 96 kHz.

Figure 5-22
Amp Farm
interface
windows

Echo Farm

Line 6
www.digidesign.com

Vintage Echo plug-ins are a collection of classic echo units. The Echo Farm collection has echo models based on Maestro Echoplexes, Roland RE-101 Space Echo, Boss DM-2, Electro-Harmonix Memoryman, and TC Electronics 2290, plus other echo effects that are all programmable and automatable (see Figure 5-23).

The maximum supported sample rate for Echo Farm plug-ins is 48 kHz.

Figure 5-23
The Line 6
EchoFarm

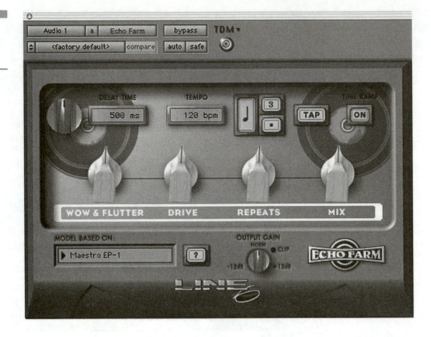

Analog Channel

McDSP
www.mcdsp.com

Analog Channel emulates the warm sounds of analog mixers and tape machines (see Figure 5-24). It has two AC1 and AC2.

AC1 controls analog gain staging, acting as a digital preamp with drive attack and release control. It is highly efficient, processing up to 24 mono channels or 16 stereo channels on a single MIX *digital signal processor* (DSP) chip. It can be used across an entire mix to create a custom analog console sound.

AC2 controls tape machine parameters such as bias, playback speed, and IEC1/2 equalization. It has adjustable low-frequency rolloff and head bump that are independent

Figure 5-24
The Analog
Channel plug-in

of playback speed. It has control for several playback head types and tape saturation from modern and vintage tape formulations.

NI Spektral Delay

Native Instruments USA
www.native-instruments.com

The NI Spektral Delay uses real-time *Fast Fourier Transformation* (FFT) to split up each channel of a stereo signal in up to 160 separately modifiable frequency bands. The level, delay time, and feedback amount for each of these bands can be set separately. Additional modulation effects can then be applied to the signal in each frequency domain, allowing significant sound manipulation.

The delay times can be set freely or according to the rhythm via a selectable note grid. The maximum delay time is 12 seconds. The effects achievable with this band delay include rather subtle colorings as well as rhythmic courses of the partial tones and dense, atmospheric sound textures.

A very interesting feature is the ability to draw the amplitudes of the bands in an Edith graph with the mouse. The filter curves resulting from this amplitude graph can have any shape and therefore make possible even very extreme filter settings and sweeps (see Figure 5-25).

Vokator

Native Instruments USA
www.native-instruments.com

Vokator is based on the principles of the classic vocoders, which are designed to combine a voice with another audio signal, creating a new sound. With Vokator, this principle can also be applied to other combinations of instruments (see Figure 5-26).

Figure 5-25
The Spektral
Delay interface

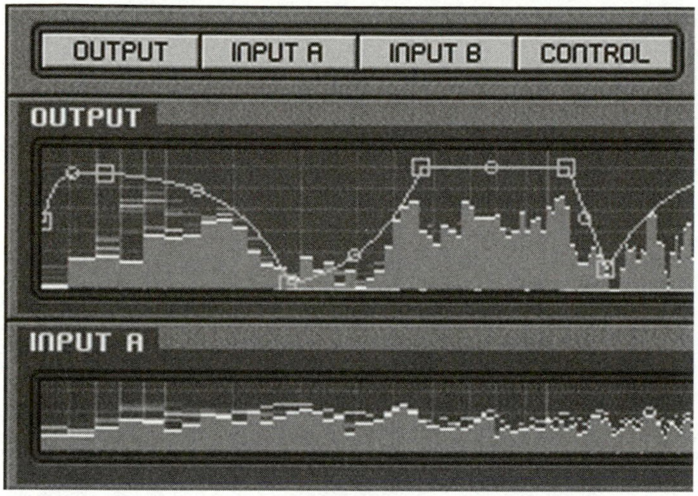

Figure 5-26
Audio signals in
Vokator

The Vokator has a library of presets that includes more than 200 total-recall presets in 10 different categories, more than 100 synthesizer/sampler snapshots in 8 categories, and 130 morph sets. It also offers 300MB samples by Zero-G in the categories Athmospheres, Bass, Drum Kit, Drum Loops, FX, guitar, harmonica, keyboard, percussion, piano, scratches, strings, synth, vocals, vocal FX, and brass (a sample can be loaded in the tapedeck or the granular sampler).

Prosoniq Orange Vocoder

Prosoniq
www.digidesign.com

The Orange Vocoder is an RTAS software synthesizer that produces a simulation of a realistic analog vocoder effect. It is fully customizable and comes with an eight-voice virtual analog synthesizer unit, Breakpoint EQ, and Filterbank Reverb (see Figure 5-27).

Figure 5-27
The Prosonic
Orange Vocoder

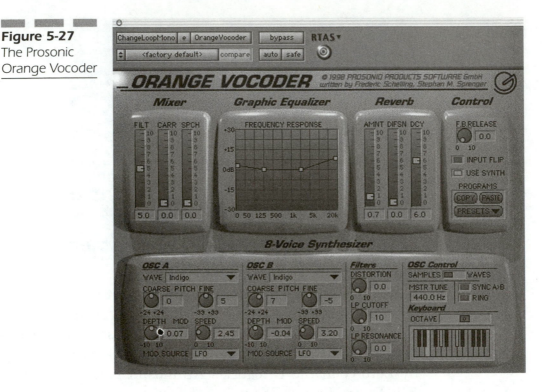

This plug-in has an integrated eight-voice virtual analog synthesizer with two oscillators per voice with 10 basic waveforms and seven sampled sounds, voice detune, pitch LFO, a four-pole low-pass filter with cutoff and resonance, oscillator hard sync, and a ring modulator.

Lexicon PSP 42

PSP Audioware
www.pspaudioware.com

Lexicon PSP 42 is a digital-stereo-delay and phrase-sampler plug-in based on the hardware Lexicon PCM 42 processor (see Figure 5-28).

Figure 5-28
The Lexicon
PSP42, a digital
version of the
PCM42

The PSP 42 generates a wide variety of delay-based effects, ranging from tempo-locked feedback delay with high-frequency absorption and tape saturation to flanging and phasing effects.

This plug-in is primarily intended for processing individual tracks within a mix. It can be used to add delay-based effects to solo instruments and vocals. It can simulate vintage delay with tape saturation, high-frequency absorption, and variable tape speed. It can produce low-fi, flanger, and doubler effects, and can act as a ring reverb. The combination of variable delay times and various internal processors can be used to create a wide range of custom effects.

Scratch

Serato Audio Research Ltd.
www.serato.com

Serato Scratch *studio edition* (SE) is a TDM, HD, and RTAS plug-in that uses an interface connected to DJs' gear to manipulate audio files by scratching with the regular DJ turntable or using a mouse (see Figure 5-29). It emulates the action and feel of a vinyl disk on a turntable with start/stop, forward/reverse pitch change. It contains a control signal that allows your computer to track the motion of the record, simulating the same movement within the digital sample.

Figure 5-29
The Scratch SE interface connects to DJs' equipment

Timeworks Delay 6022

Sonic Timeworks
www.sonictimeworks.com

The Timeworks Delay 6022 has independent delay time, feedback gain, and delay gain for each channel. It has delay times up to 1.5 seconds, tone controls for each channel, and switchable metering of input/output or delay levels. The plug-in also includes high-frequency damping

Figure 5-30
The Timeworks
Delay window

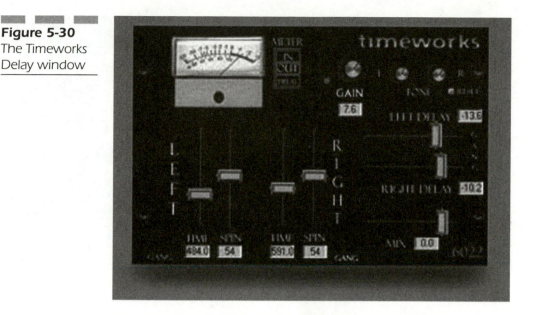

Figure 5-30
The Timeworks
Delay window

on the delayed part, which is independent for each channel and fader, ganging up to facilitate adjustment of both channels simultaneously. It works on PCs only (see Figure 5-30).

Timeworks Phazer Model 88

Sonic Timeworks
www.sonictimeworks.com

The Model 88 sounds like an old-style foot-pedal phaser. It's a full-bodied phaser with zero noise, broad frequency response, adjustable center frequency, and 64-bit internal precision. It's designed for quick and precise adjustments. It has mono and stereo input and output with mono-stereo conversions, and an oscilloscope, which displays music in colors (see Figure 5-31).

Figure 5-31
The Phazer
Model 88
interface, which
includes an
oscilloscope

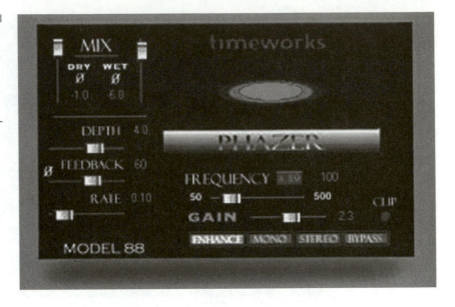

FilterFreak

SoundToys
www.soundtoys.com

SoundToys are a new generation of plug-ins. The first
SoundToys plug-in is FilterFreak, an analog-modeling filter plug-in that combines analog filter modeling and creative DSP design (see Figure 5-32). Based on studio
filtering boxes like the Mutronics Mutator and the Sher-
man FilterBank, it is designed for radical filtering effects.

The FilterFreak has a 48 dB per octave analog-modeling
filter with adjustable filter slope, resonance, and filter
shape. It has five different modulation sources, a tempo-
locked LFO, envelope follower, random LFO, ADSR, and a

Figure 5-32
The first Sound
Toys plug-in

triggered random stepper. It also has user programmable wave shapes and programmable rhythmic patterns.

Extension 78

Summit Audio
www.summitaudio.com

Extension 78 is a TDM plug-in that controls up to 32 channels of Element 78 external hardware preamps and EQs via a MIDI interface.

The Element 78 series hardware includes two MIDI-controllable microphone preamps with built-in EQ, the MPE-200 and the MPE-200S. The MPE-200 has a full complement of knobs and buttons for front panel control, while the MPE-200S is optimized for remote-control use with just one knob and a button on its face (see Figure 5-33). There are also two two-channel parametric equalizers, the EQ-200 with front panel controls, and EQ-200S without. Each hardware unit costs several thousand dollars and requires a dedicated MIDI port.

Figure 5-33
Extension 78

PS22 StereoMaker

Waves
www.waves.com

The PS22 StereoMaker is designed as a dedicated stereo-from-mono tool. It creates a stereo image from mono sources. It also works with stereo inputs to enhance the spread of stereo sources. It can be used on single tracks of a mix, such as lead instruments or backing vocals, or for stereo enhancement of full mixes or old recordings.

The StereoMaker has very low coloration, high tonal accuracy, and good mono compatibility. The interface provides controls for repositioning specific sounds in the stereo image, and a graphical display of position as a function of frequency (see Figure 5-34).

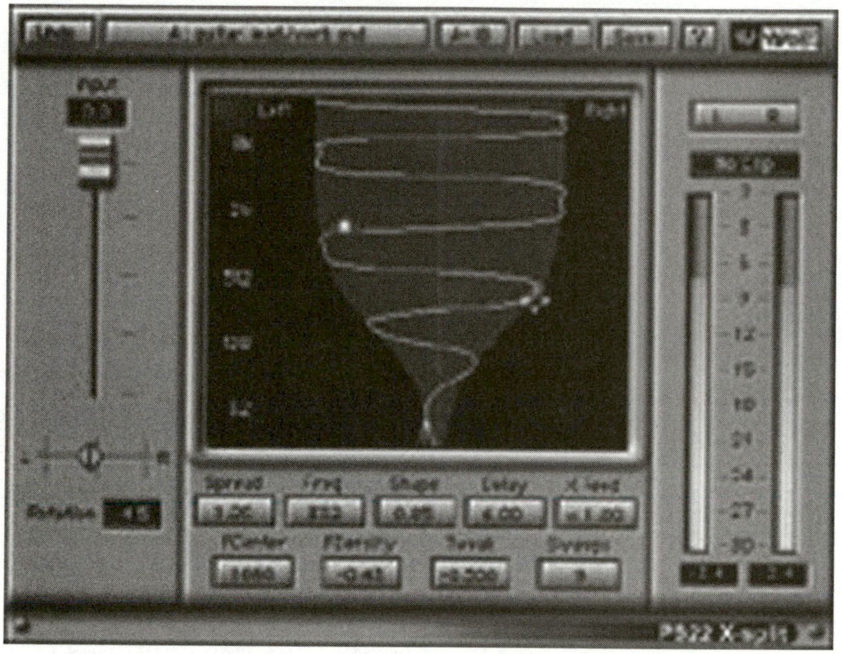

Figure 5-34
Frequency
displayed in the
StereoMaker
interface

Figure 5-34
Frequency
displayed in the
StereoMaker
interface

SoundBlender

Wave Mechanics
www.wavemechanics.com

SoundBlender includes two plug-ins, PitchBlender and TimeBlender, that can be combined to create a diverse collection of time, pitch, filter and modulation-based effects, all available simultaneously and in real time. It has two channels of intelligent pitch-shifting, delay, panning, amplitude modulation, and resonant filtering. There is also a powerful modulation matrix allowing users to mix any variation of the modules in real time (see Figure 5-35).

The plug-in comes with over 200 preset patches, including chorus, delay, filter, pitch shift, and panning effects, as well as numerous extreme effects.

The PitchBlender has two channels of very high quality conventional pitch shifting with a shift range of +/–2 octaves. The pitch-shifters are optimized for small-interval pitch shifting for clean chorus and double-track effects. The output level and panning of each pitch shifter may be adjusted by onscreen faders and smoothly modulated by the modulation matrix.

Each pitch shifter can use intelligent pitch processing for diatonic interval transposition. Any modulation source may be used to control the pitch-shift interval to create in-key arpeggiation and randomized harmony effects. A wide variety of western and ethnic scales are available for a wide range of pitch textures. Each channel has up to 650 milliseconds of delay available. The output level and panning of each delay line may be adjusted by onscreen faders and modulated by the modulation matrix. Each filter may be configured as a low-pass, high-pass, band-pass, or notch filter. Each filter has adjustable center frequency and resonance control. Center frequency can be modulated by the modulation matrix.

The TimeBlender module has two channels of reverse pitch shifting. Each channel can sample segments as long as 1,000 milliseconds, and can vary the playback pitch of each segment over a two-octave range.

A master *beats per minute* (BPM) control can be used to synchronize delay and modulation effects to the tempo of a Pro Tools session.

Multifunction Effects Processors

Multifunction effects processors combine the functions of regular effects processors. They allow you to create more effects with fewer plug-ins.

Pluggo 3

Cycling 74
www.cycling74.com

Pluggo 3 is a real-time interactive audio-processing, modulation, and synthesis environment that works as a plug-in. It is a collection of more than 100 audio plug-ins based on its own processing system, the Max/MSP audio processing system and its associated plug-in development tools.

When being used in Pro Tools (or other sequencers), Pluggo uses a run-time version of Max/MSP, which all the plug-ins run within (see Figure 5-36). The individual plug-

Figure 5-36
One of Pluggo
3's modules

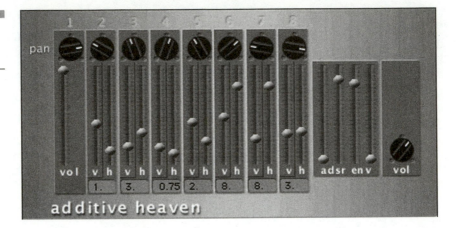

ins communicate with each other directly in Max/MSP in ways that are set up by the user. For example, the Pluggo-Sync plug-in can be set up so that it listens to an audio click track, and then tells other plug-ins where the beat is. Pluggo plug-ins can send and receive MIDI information. Also, the PluggoBus system lets plug-ins send up to eight channels of audio to each other.

Effects categories include delays, filters, pitch effects, distortion, granulation, spectral effects, modulators, multichannel effects, synthesizers, audio routing, reverb and dynamics, and visual display. There are also 19 instrument plug-ins that were created by eowave, the developers of the iSynth. These include the Pretty Good Synth, additive synths, analog-modeling drum and percussion synths, theremins, sampling, granular synthesis, FM synthesis, wave-table synthesis, and wave-shaping tools.

D-Fi

Digidesign
www.digidesign.com

D-Fi is a family of four unique, creative sound-design plug-ins (Lo-Fi, Sci-Fi, Recti-Fi, and Vari-Fi), providing tools for the TDM, RTAS, and audiosuite environments (see Figure 5-37). Lo-Fi provides bit-reduction for retro sound

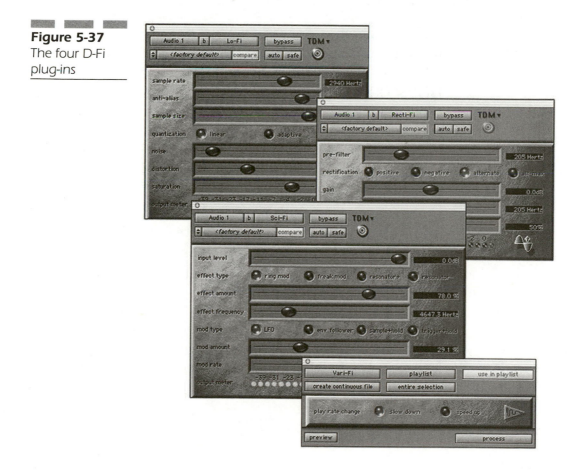

Figure 5-37
The four D-Fi plug-ins

processing without the expense of retro equipment. Sci-Fi adds analog, synth-type ring modulation, frequency modulation, and variable frequency resonators. Recti-Fi provides super- and subharmonic synthesis, and Vari-Fi allows the processing of disk files to create the effect of audio changing speed to or from a complete stop. The maximum supported sample rate is 192 kHz.

DSPider

DUY Research
www.duy.com

DUY's DSPider is a TDM-based, modular processing software platform that consists of 40 separate modules, which can be flexibly linked by the user to create new user-defined processors (see Figure 5-38). It operates in two basic modes. In Advanced mode, users can create or edit patches, allowing the development of unique proprietary processors and signature sounds for a broad range of applications, including post-production, music, broadcast, mastering, and multimedia. Reader mode is a simplified interface used when applying the effects.

The plug-in comes with a library of over 220 presets, including compressors, reverbs, equalizers, limiters, synths, noise reduction systems, 3D effects, de-essers, and sound effects generators. New patches will be released and updated periodically by DUY.

Figure 5-38

DSPider's link-based interface

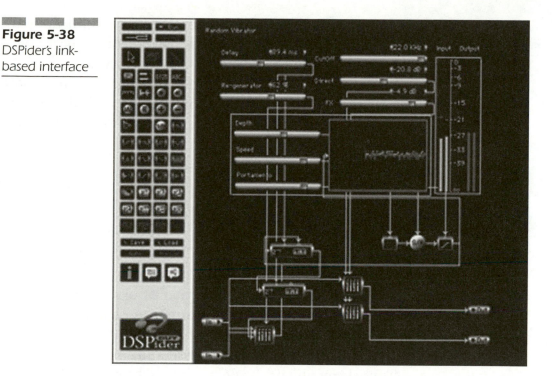

DUY Shape

DUY Research
www.duy.com

DUY Shape is a wave shaper based on DUY's *frequency-dependent wave shaping* (FDWS) algorithm. It has a three-band smooth filter with full audio range, continuous crossover points, and analog sound filters that use modeling technology.

There are three independent, user-defined shapers with eight different shaper curve types, including linear, log, and cosine functions. It has simultaneous input and output plasma-like meters. It can be used for the processing of independent tracks or mastering final mixes.

Shape can be used as a dynamic enhancer, a smooth equalizer, or a wave-shaping compressor, and also for frequency enhancing and frequency redistribution with user-defined parameters (see Figure 5-39).

Epic TDM

emagic
www.emagic.de

The Epic plug-ins are a collection of eight plug-ins that were derived from Logic Platinum (see Figure 5-40). They include Tape delay, Tremolo, Autofilter, Ensemble, Spectral Gate, ES1, Enveloper, Phaser, and subBass.

Tape delay allows the delay time to be set in note values with an adjustable groove. The feedback path has independent high- and low-pass filters, and there are multiple parameters for the simulation of the characteristics of original analog devices. This version offers an additional LFO and is available in stereo.

Figure 5-39
DUY Shape

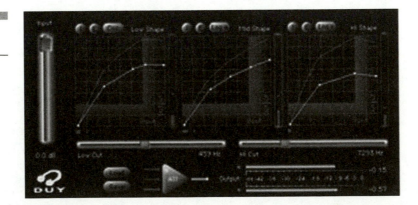

Figure 5-40
The Epic TDM
plug-in suite

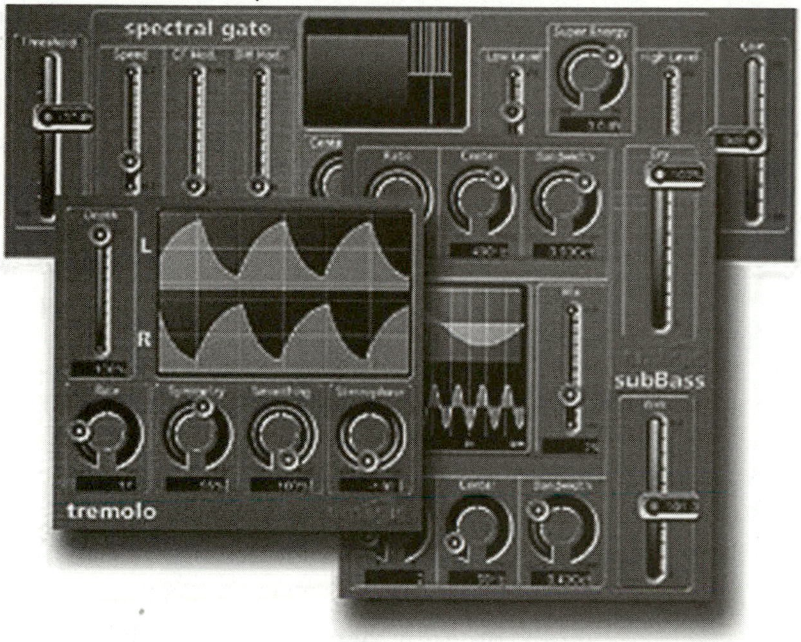

Tremolo has adjustable depth, rate, symmetry, smoothing, and stereo phase.

Autofilter is a 6, 12, 18, or 24 dB per octive low-pass filter with adjustable fatness for the preservation of low frequencies with high resonance. It features a level-triggered envelope, an LFO, and pre- and postdistortion circuits.

Ensemble creates modulation effects with up to eight voices, using three mixable LFOs for pitch modulation. It can transform single sawtooth waves into an eight-voice ensemble.

Spectral Gate is an FFT-based effect that allows a frequency and level selective isolation of signal components.

ES1 is an analog-style synthesizer with flexible tone generation with a superb sounding filter section.

Enveloper is designed to control and individually emphasize or attenuate attack and release phases (transients or reverb tails) of any signal.

Phaser provides 12 phase-shift stages with feedback and two mixable LFOs, whose modulation depth can be adjusted as a lower- and upper-notch frequency.

The subBass plug-in is used to add deep base frequencies to any signal.

GRM Tools' Classic RTAS and TDM

INA–GRM
www.grmtools.org

GRM Tools' Classic TDM and Classic RTAS are bundles of eight plug-ins each that provide superb tools for sound enhancement and design. Conceived and realized by the Groupe de Recherches Musicales (Musical Research Group) of the National Audiovisual Institute, Paris, France, GRM Tools is the result of more than 20 years of research and development by composers and sound designers in sound transformation software.

Seven of the eight plug-ins are similar when in the TDM and RTAS sets. These include the Comb Filters, Delays, Doppler, Band Pass, Freeze, Shuffling, and Pitch Accum (see Figure 5-41). In addition, the RTAS version includes Reson and the TDM version includes Equalizer.

Comb Filters RTAS and TDM

This plug-in gives you five high-Q in-parallel comb filters, controllable independently or in ensemble, plus five low-

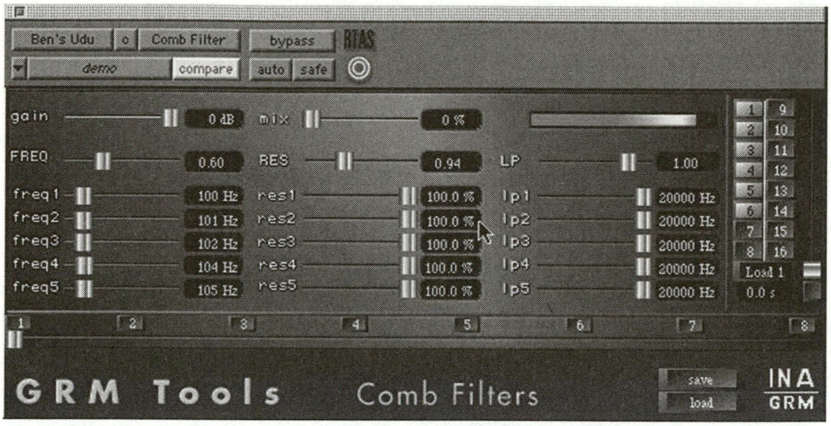

Figure 5-41
The Classic RTAS
and TDM

pass filters to control the high-frequency content of the output.

Sliders let you control individual filter frequencies, the individual resonances of each of the filters, and the cutoff frequencies of the five low-pass filters respectively. The FREQ, RES, and LP sliders let you control the frequencies, resonances, and low-pass cutoffs together.

Delays for RTAS and TDM

The delays can be used to create any type of delay line, including echo, reverberation, and phase shift, as well as many subtle effects.

This plug-in provides up to 128 delays within a range of up to 6 seconds. The user has control of their individual amplitudes and timing, allowing many possibilities in a custom amplitude curve and in the resulting sound. In the stereo version of this plug-in, the delays are assigned alternatively to the left and right channels.

Doppler RTAS and TDM

This plug-in lets you simulate the doppler effect, which is your perception that a sound moving toward you rises in pitch while a sound moving away from you falls in pitch.

Band Pass RTAS and TDM

This plug-in is a dynamically controllable band pass filter capable of targeting frequencies for extreme equalization. The 2D Controller in the center of the control window can be dragged to shape the filter curve.

Freezing RTAS and Freeze TDM

As a sound passes through the Waveform window, a click will freeze a three-second segment of the sound in the window and it will be played continuously as a number of simultaneous loops. The loops can be varied in duration and pitch to encompass the whole or any part of the 3-second segment.

Shuffling RTAS and TDM

Shuffling can be used to create reverb, echo, and spatialization effects. It fragments a sound and then allows you to vary the pitches of the fragments and randomly shuffle them in time.

Pitch Accum RTAS and TDM

Pitch Accum has two independent and simultaneous transposers that can be used with delays and modulations to transpose the pitch of an input sound, while a pitch follower detects the pitch of the input sound and optimizes the transposition. It lets you create two distinct shadows of a sound at different transposition levels and delay intervals, and then modulate those sounds in different ways.

Reson RTAS

Using format-synthesis technology, this plug-in generates up to 128 resonating high-pass, low-pass, or band-reject filters, each of which resonates a single frequency and allows you to distribute them within a defined frequency range.

Equalize

Equalize is used to rebalance the spectrum of a sound by changing the gain of different frequency bands independently or together.

GRM Tools ST

INA–GRM
www.grmtools.org

The *Spectral Transform* (ST) bundle is a collection of four plug-ins, including Contrast ST, Equalize ST, Shift ST, and FreqWarp ST.

Contrast ST is used to add vibrancy, liveliness, depth, and subtle changes in timbre to sounds in mastering applications. It can also be used to create a wide range of extreme transformations and effects (see Figure 5-42).

Contrast analyzes a user-defined frequency range within the spectrum of an input sound and groups the frequency components within that range according to the strength of their amplitudes. The groups are frequency components with strong amplitudes, frequency components with medium amplitudes, and frequency components with weak amplitudes. The strength of each of the groups can then be modified independently.

This approach allows you to redefine the character of a sound. If you decrease the strength of the medium and weak amplitudes, thus emphasizing the strong amplitudes, for example, you'll make a sound more mellow. If you strengthen the medium amplitudes relative to the stronger and weaker amplitudes, you'll add vibrancy, live-

Figure 5-42
The Contrast ST
interface

liness, and depth to a sound. If you strengthen the weak amplitudes relative to the others, you'll make the sound more metallic, harsh, and percussive.

Equalize ST is a 31-band graphic equalizer with dynamic controls to rebalance the highs, lows, and mid-range of a sound. In the stereo version of this plug-in, you can control the equalization curves of left and right channels independently or together. Each band, controlled by a single slider, is $1/3$-octave.

Shift ST contains two functions that can be used separately or together. The scale function transposes a sound by multiplying each spectral component by a constant frequency. The shift function is a frequency shifter, sometimes referred to as a single-sideband ring modulator, that adds a constant frequency to each spectral component. When you use a frequency shifter, harmonic sounds become inharmonic. This plug-in can be used to transpose or transform a sound by any combination of frequency scaling or frequency shifting.

FreqWarp ST is used to transform a sound by rearranging its frequency components with a unique graphical interface. The frequencies of a sound are represented as points on a diagonal line with the y-axis representing output and the x-axis representing input.

When you click on the diagonal line, or anywhere in the Control window, you create a junction point at a specific position that represents a source frequency in the input sound and a destination frequency in the output sound. You are, in effect, transferring a frequency from the input sound to a new position in the output sound, and the output sound will contain a rearrangement of the frequency components of the input sound. This can result in a radical change in timbre.

ChannelStrip/ChannelStrip SP

Metric Halo Distribution Inc.
www.mhlabs.com

ChannelStrip has exceptional audio quality, high DSP and *central processing unit* (CPU) efficiency, and a comprehensive user interface (see Figure 5-43). It has an expander, a gate, and a compressor with a side-chain filter, a six-band parametric EQ, a phase invert switch, and a sample delay parameter. There are virtual knobs and buttons for adjusting parameter settings. There are also three peak meters for each module and a gain-reduction meter for the compressor. Five adjustable response graphs provide a visual display of each module and the side chain inputs. In the center there are stereo meters and a gain control.

The dynamics section of ChannelStrip offers the choice of processing the signal pre- or post-EQ. It has three different modes of compression. Warm is the most versatile setting. Fast is good for quick transients. Smooth is useful

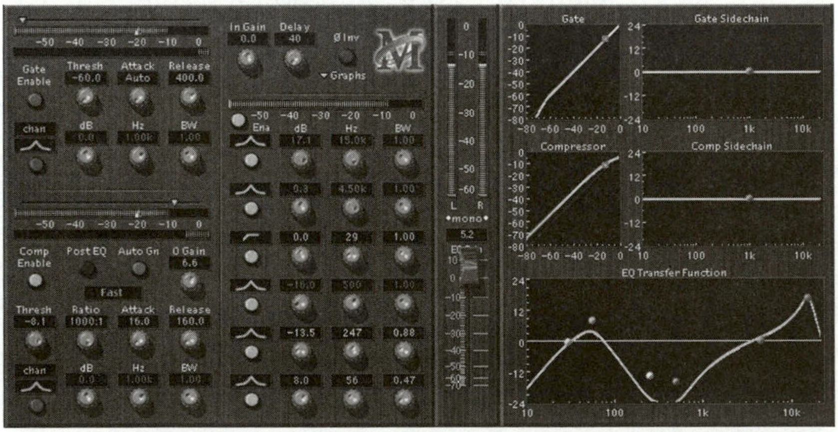

Figure 5-43
The Channel-
Strip UI

for mix-bus compression. There are standard compression controls for adjusting threshold, attack, release, and ratio. There is also an automatic gain control feature.

The EQ module provides six bands of 48-bit parametric equalization (eight if you include the two side-chain bands). Each band has six different types of filter shapes, including peaking/parametric, high-shelf, low-shelf, high-cut, low-cut, and band-pass.

ChannelStrip ships with 127 presets in 11 different categories with settings for full mix, vocals, drums, and other instruments.

ChannelStrip SP is a less expensive, "lite" version that has fewer functions. It provides the same audio quality as the full version.

TC Tools 3.6

TC Works
www.tcworks.de

TC Tools 3.6 from TC Works includes reverb, chorus, and parametric EQ plug-ins.

The reverb plug-in, Mega-Reverb, is based on the core technology of the TC Electronic M5000 high-end studio processor. It uses the reverb algorithms Core 1 and 2 from the TC M5000, with new diffusion and tail tuning. It includes six modeled room shapes: hall, fan, prism, horse-shoe, small, and club. In addition, it has separate early reflections and tail controls, a high-cut filter, stereo processing, 100 presets, and 24-bit 96 kHz processing (see Figure 5-44).

The chorus/delay plug-in recreates the 1210 Spatial Expander with its unique modulation images like chorus,

Figure 5-44
TC Works; TC
Tools 3.6

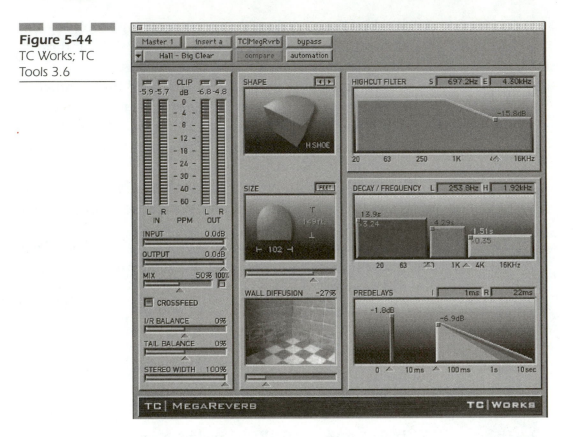

flanging, and slap delay, with very flexible routing and filtering facilities, which can be adjusted in BPM. It has 10
chorus, flanging, and slap echo delay effects.

EQsat is a very clean-sounding and carefully crafted EQ
model with three parametric bands and two additional
shelving bands. The frequency response display gives
instant feedback on the applied equalization curve. The
plug-in also has a saturation emulation for an analog
sound.

AudioTrack

Waves

www.waves.com

AudioTrack is a channel strip plug-in with a parametric four-band EQ, with bell, shelf, and high-pass/low-pass filters, modeled after the Q10 equalizer. It also has a compressor and a noise gate. There are many presets for single-pass EQ, compression, expansion, and gating.

The single-window interface has volume and gain reduction meters; sample-accurate peak hold and clip meters; instantaneous A/B comparisons between online settings; and a setup library with effect setups, multimedia setups, musical style setups, RealAudio setups, and ShockWave audio setups (see Figure 5-45).

Figure 5-45
AudioTrack's
single-window
interface

C1 Parmetric Compander

Waves
www.waves.com/

The C1 Parametric Compander is a one-band dynamic equalizer and frequency-selective dynamic processor with compression, expansion, limiting, and gating capabilities (see Figure 5-46). It also includes phase compensation for transparency and quality.

The C1 allows two simultaneous dynamic processes in any frequency range and can also be used as a traditional wideband dynamics device.

Figure 5-46
The CI
Parametric
Compander

C4 Multiband Parametric Processor

Waves

www.waves.com

The C4 Multiband Parametric Processor is a four-channel version of the C1 (see Figure 5-47). It is based on the Renaissance Compressor design with a crossover designed with ideal phase characteristics to achieve powerful control while remaining transparent.

Different functions in each band can be performed simultaneously, all with continual visual feedback.

The plug-in has four fully parametric bands for any combination of compression, expansion, limiting, or EQ.

There is an independent threshold, range, gain, attack, release, and bandwidth for every band. The display shows all parameters simultaneously.

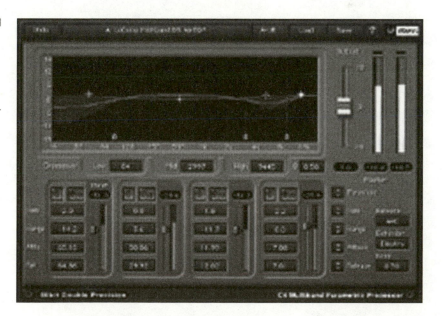

Figure 5-47

The expanded
C4 Multiband
Parametric
Processor

CHAPTER **6**

Making Quality
Audio

Pitch/Tempo/Voice Processors

Several plug-ins available for Pro Tools can be used to affect the basics of the music pitch and tempo. Processors also exist that allow you to affect voicing, range, and style. All of these plug-ins can be used to improve the quality of your audio project and to create the final results that you want.

Auto-Tune 3

Antares Audio Technologies
www.antarestech.com

Auto-Tune 3 is a powerful pitch-correction plug-in that corrects intonation problems of vocals and solo instruments in real time.

The graphical mode displays the detected pitch and allows you to draw in the desired pitch on a graph (see Figure 6-1). This mode gives complete control over adding or repairing scooping pitches and large vibratos.

The automatic mode instantaneously detects the pitch of the input, identifies the closest pitch in a user-specified scale, and corrects the input pitch to match the scale pitch. Major, minor, chromatic, and a collection of historical and microtonal scales are included. You can also input custom scales.

The plug-in has source-specific pitch detection and pitch-correction algorithms with settings that include soprano voice, alto/tenor voice, low male voice, instrument, and bass instrument. Matching the appropriate algorithm to the input results in even faster and more accurate pitch

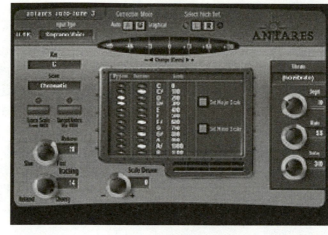

Figure 6-1
Auto-Tune 3's
Automatic Mode

Corrects the pitch of a vocal or solo instrument in real time, without distortion or artifacts, while preserving all the expressive nuance of the original performance.

detection and correction. It provides phase-coherent pitch correction of stereo tracks.

Auto-Tune also has a special Bass mode that lowers the lowest detectable frequency by about one octave to 25 Hz. Since the lowest E string on a bass guitar is approximately 41 Hz, Bass mode allows the user to apply pitch correction to fretless bass lines as well as other low-bass-range instruments.

The plug-in functions at high sample rates and has 88.2 kHz and 96 kHz compatibility with Pro Tools.

JVP (Voice Processor)

Antares Audio Technologies
www.antarestech.com

JVP is a software plug-in that incorporates a de-esser, a compressor with a downward expanding gate, a parametric EQ, and a multitap delay in a single interface. It can simultaneously de-ess, compress, gate, EQ, and add delay effects to a vocal or solo instrument track, or master a stereo mix.

The Parametric EQ has three bands, five filter types, and a true 140 dB dynamic range, making this plug-in one of the best-sounding EQs available anywhere. The compressor/gate has a variable knee and special compression algorithms to provide smooth, natural dynamic control. The de-esser is fully programmable and easy to use (see Figure 6-2). The multitap stereo delay has six taps with feedback for complex time-based effects. JVP processes up to 24-bit files with an internal 56-bit accumulator for excellent stereo or mono sound quality.

H949

Eventide
www.eventide.com

Eventide's H949 is the software version of a classic hardware pitch changer (see Figure 6-3). The hardware version of the H949 was made from 1977 to 1984 and was the first

Figure 6-2
The easy-to-use
JVP

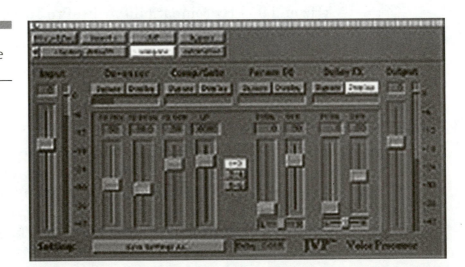

Figure 6-3
Eventide's digital
version of the
H949

deglitched pitch changer. It was hugely popular in studios during the 1970s and 1980s, and has a unique sound.

Pitch 'n Time

Serato Audio Research Ltd.
www.serato.com

Serato Pitch 'n Time is a time-stretching and pitch-shifting audiosuite plug-in for Pro Tools. It provides time compression and expansion from a ratio of 1/8 speed through to 8x speed, independent of pitch. It also includes a pitch-shifting capability of up to 36 semitones, independent of tempo. The time-stretching and pitch-shifting functions each have three interchangeable panels, ranging from simple, fixed ratios to complex tempo and pitch alterations that vary over time. This is a well-developed product and provides clean functionality (see Figure 6-4).

VocALign Project

Synchro Arts Limited
www.synchroarts.com

The VocALign Project is an entry-level version of VocALign Pro that is designed as a plug-in for Pro tools

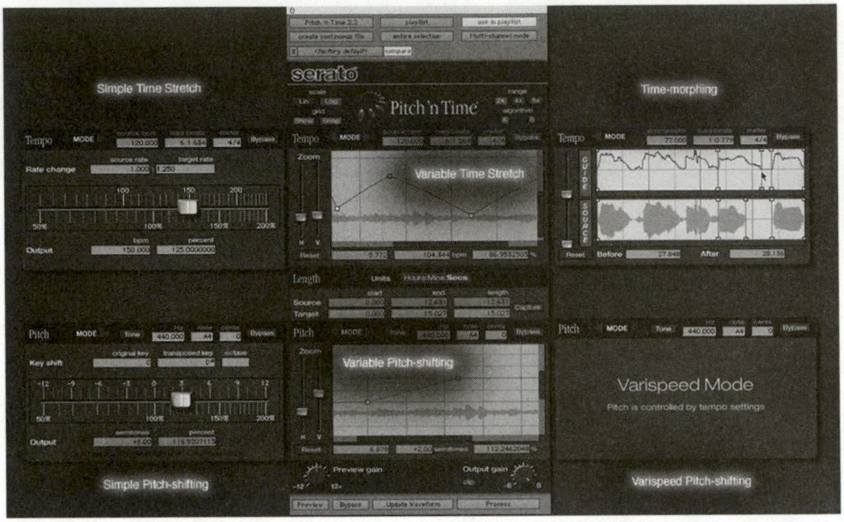

Figure 6-4
Serators Pitch 'n
Time

(see Figure 6-5). It is an editing tool that will automatically synchronize two audio signals at the touch of a button. It dramatically speeds up audio synchronization tasks and increases the quality of results.

VocALign is useful for lead vocal doubles, backing vocals, instrumental tracks, and *automatic dialog replacement* (ADR) where it will automatically edit a line of replacement dialog so that it aligns with the dialog recorded with the original film or video, creating perfect lip-sync and more convincing foreign language dubs.

PitchDoctor

Wave Mechanics
www.wavemechanics.com

PitchDoctor is a pitch-shifting plug-in. It combines an intelligent pitch analyzer and a pitch shifter to provide real-time pitch correction. PitchDoctor automatically

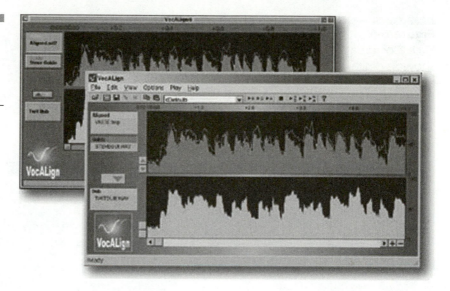

Figure 6-5
VocAlLign
synchronizes
tracks
automatically.

adjusts the intonation of any out-of-tune notes to the user-defined key and scale.

The onscreen keyboard displays the musical scale used for pitch correction, allowing easy scale customization while also providing a visual readout of the detected note (see Figure 6-6). Since PitchDoctor is fully automatable, the key, scale, and scale customizations may be changed in midtrack.

With the score control feature, Pitch Doctor will smoothly modify a performance to bring it into alignment with a specified melody. With the score generation feature, any instrument can be used to create a score to guide PitchDoctor through the pitch-correction process. There is also a manual-correction fader that can either be used alone or in conjunction with the automatic pitch-correction mode.

Figure 6-6
Pitch Doctor
shows a digital
readout of
notes.

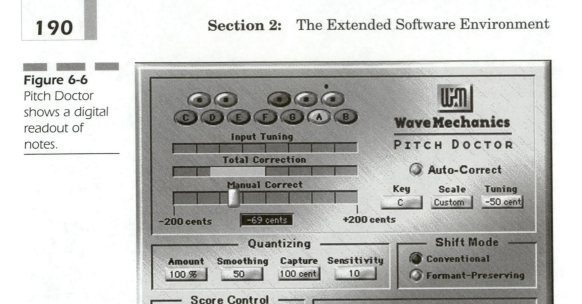

Figure 6-6
Pitch Doctor shows a digital readout of notes.

PurePitch

Wave Mechanics
www.wavemechanics.com

PurePitch is a pitch shifter that can perform some interesting tricks, including extreme vocal-processing effects and the creation of harmony parts from a lead vocal or instrument (see Figure 6-7). It can deepen a vocal part or add vibrato without changing its pitch. It can be used to alter the formats and pitch inflections of a speaker, producing subtle character alterations, or more extreme morphing effects. There are over 50 preset patches for music production, sound design, and post-production.

Figure 6-7
The Pure Pitch
interface

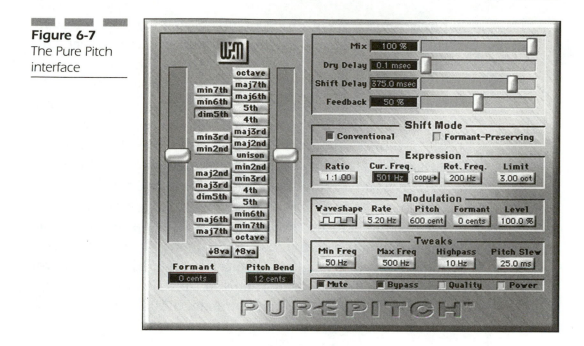

Speed

Wave Mechanics
www.wavemechanics.com

Speed is a simple pitch and time processor in the form of a well-integrated audiosuite plug-in.

An audio range is first selected in the Pro Tools Edit window and then Speed is chosen from the AudioSuite menu. The Preview button at the bottom of the window lets you audition the current tempo and pitch settings (the default settings signify no change to your audio). You use the Tempo Control knob to change tempo without altering pitch, and the Pitch Control knob to change pitch without altering tempo. You can also click on the displayed value and type in a new setting. Both the speed control and the pitch control have different modes, which you select by clicking on the small, white, triangular Mode Select tabs in the middle of the window (see Figure 6-8).

The tempo control has two modes: tempo and length. A tempo setting above 100 percent increases the tempo, and a setting below 100 percent reduces the tempo. For example, a tempo setting of 200 percent will result in a tempo twice as fast as the original, and a setting of 50 percent will give you a tempo half as fast. (Settings for tempo and length range from 50 to 200 percent, with a resolution of six decimal places in Simple and Calculator mode and three decimal places in Graphical mode.) In Length mode, higher settings lengthen a selection, resulting in a slower tempo, and lower settings shorten the selection, creating a faster tempo. As with Tempo mode, a 100 percent length setting has no effect on the audio.

Figure 6-8
Speed, by Wave
Mechanics

Pitch control has three modes: cents, semitones, and percent. Both cents and semitones let you shift your audio a maximum of one octave in either direction. Percent mode uses a frequency ratio to define the change; a value of 100 percent, for example, won't modify the pitch. (Settings range from 50 to 200 percent.) This mode resembles the pitch-ratio control of older hardware pitch-change devices and can compensate for the pitch change produced by sample-rate conversion.

There are also controls for the numerical input of values and a graphical display that allows manipulation of the sound via the movement of points on a graph.

Noise Reduction

DINR

Digidesign
www.digidesign.com/

Digidesign Intelligent Noise Reduction (DINR) is an intelligent broadband noise-reduction and hum-removal plug-in that works by finding and subtracting noise from digital audio files. Applications include the elimination of tape hiss, guitar-amp buzz, air conditioner rumble, and other unwanted leakage and artifacts.

The noise signature is created by selecting and analyzing an example of the noise within the source material. Then an onscreen slider can be used to apply from 0.1 dB to an infinite amount of noise reduction. Other parameter settings can also be adjusted on the fly to maximize noise reduction while minimizing signal loss and artifact generation. The resulting audio has minimal, if any, distortion, dynamic modulation, or fluctuations in frequency response.

Parameter settings can be saved and recalled for use on other source-audio files affected with similar noise. Additionally, the noise signature of any device can be kept and used with other source material the device may affect. Noise signatures and parameter settings are also automatically stored with the Pro Tools session for instant recall.

NoNOISE

Sonic Solutions
www.digidesign.com

NoNOISE is a set of powerful noise-reduction and audio-restoration plug-ins that address the full spectrum of noise issues (see Figure 6-9). It is a plug-in version of the Sonic Solutions Sonic Studio product, which is used heavily in professional restoration and is quite expensive. It has a variety of high-resolution filters for removing artifacts such as hum and buzz, and for applying standard filter types commonly used in mastering, including presence, notch, high and low shelf, band pass and band stop, DC-removal, emphasis, deemphasis, and RIAA/No RIAA.

The plug-in includes tools for frequency analysis, peak distortion removal, broadband noise reduction, click detection, production (automatic) declicking, and manual declicking, and decrackling. It can be used for the mastering and remastering stages of audio production, and for many other audio treatments needed to restore, clean, or enhance recorded material. Its processes isolate and elim-

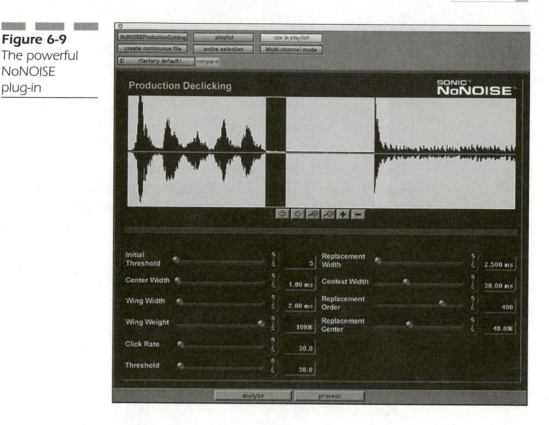

Figure 6-9
The powerful
NoNOISE
plug-in

inate audio artifacts such as hiss, scratches, hum, and mechanical and impulsive noise, without damage to program material.

Mastering Tools

Mastering in Pro Tools is made possible by the array of available mastering plug-ins. If you are interested in mastering your projects, these plug-ins can enhance your system.

MasterTools

Apogee
www.apogeedigital.com

When the MasterTools UV22 Mastering Software was first released in 1993, the UV1000 Super CD Encoder cost almost $7,000. It rapidly became indispensable to the mastering community. It is still a very popular system for mastering high-resolution digital signals to CD.

MasterTools with UV22 is designed to retain maximum detail for CD mastering. It has a 3-D multifunctional digital meter display, customizable "over" (illegal signal) indication and logging, DC removal, and NOVA, which ensures that no "overs" reach the final master.

The overs are logged with timecode so that the user can return to the mix and compensate. Another alternative is to use the NOVA feature, which automatically eliminates overs.

Both peak and average audio levels are displayed. The peak hold times and meter scaling are both user-definable. Analog and digital scales with selectable headroom are also available.

The plug-in also includes phase invert, channel reverse, and mono settings that can be used to check phase and stereo image. There is also a comprehensive DC-offset removal system in which the display shows the amount of DC offset that is being corrected in real time (see Figure 6-10).

Figure 6-10
The MasterTools
interface

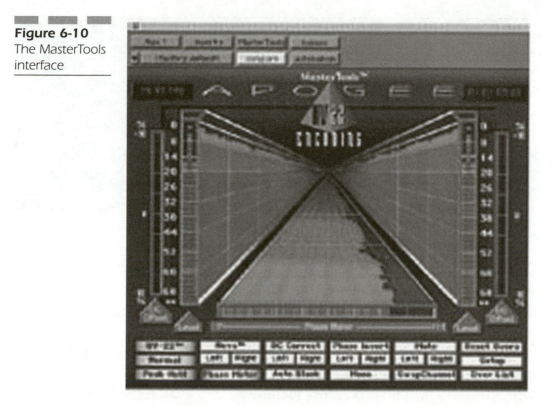

DUY Max

DUY Research
www.duy.com

Max DUY is a sound-level maximizer. It uses a proprietary
intelligent-level-optimization (ILO) algorithm. It provides
seamless level maximizing with no harmonic distortion,
no unnecessary level scaling and limiting, and release-free
operation. This software delivers a noticeable difference in
the signal-to-noise ratio with no unwanted pumping
or other artifacts. The interface is intuitive (see Figure 6-11). Applications include CD mastering, audio post,
and multimedia.

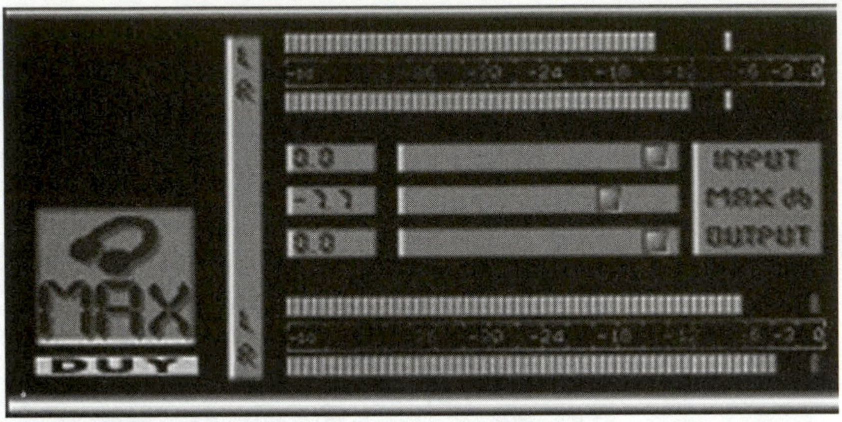

Figure 6-11
DUY Max

T-RackS

IK Multimedia
www.t-racks.com

T-RackS is a multistage plug-in with a multiunit rack paradigm in the interface (see Figure 6-12).

The processors include an EQ, a tube-modeled compressor, a multiband master limiter, a soft-clipping output stage, and a complete mastering suite. The plug-in enhances mix frequencies, stereo images, and dynamic range, and gives your audio a seamless, top-notch sound.

The unit is designed to provide a warm tube compressor sound in a full mastering suite.

Master X3

TC Works
www.tcworks.de

Master X3 is the plug-in version of the TC Electronic Finalizer, a high-end hardware all-in-one mastering device

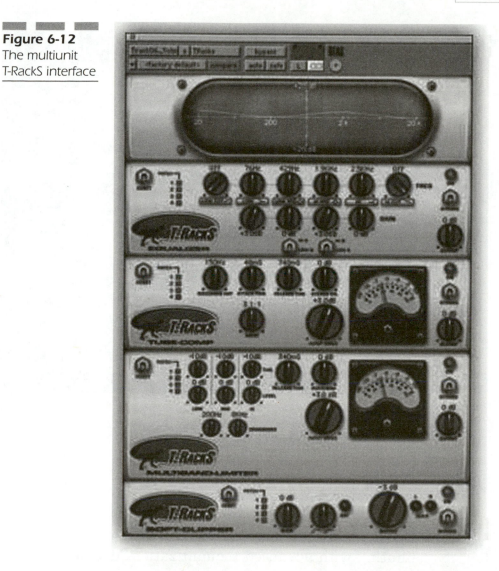

Figure 6-12
The multiunit
T-RackS interface

(see Figure 6-13). Its functions include shaping, enhancing, normalizing, maximizing, expansion, compression, and limiting. It also performs high-resolution fades and dithers. It can operate at sample rates up to 96 kHz.

Figure 6-13
Master X3, the plug-in version of the TC Electronic Finalizer

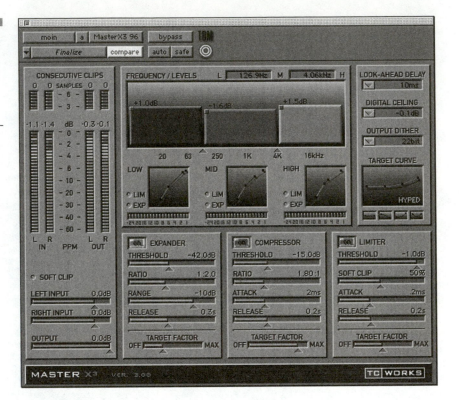

L1 Ultramaximizer

Waves
www.waves.com

L1 Ultramaximizer is a maximizer with look-ahead peak limiting with advanced requantization. It is designed to provide maximum levels for CD mastering, multimedia, and other mastering applications. It can also be used as a mono or stereo limiter. It processes with 48-bit double precision, dithered to 24-bit output. It has three noise-shaping curves, sample-accurate peak-hold metering, and a spacy adjustable release time (see Figure 6-14).

Figure 6-14
The L1
Ultramaximizer

L2 Ultramaximizer

Waves

www.waves.com

The L2 Ultramaximizer is a software version of the L2 hardware processor (see Figure 6-15). It is designed to increase the average signal level of audio signals without introducing audible side effects. It has a brick wall look-ahead peak-limiter algorithm and Waves *increased digital resolution* (IDR) dithering technology with ninth-order noise-shaping. It uses 48-bit internal processing and outputs at 24, 22, 20, 18, and 16 bits.

In addition to mastering, the L2 Ultramaximizer can be used for mixing and recording, and as an extreme limiter or for dynamic processing effects.

Figure 6-15
The L2
Ultramaximizer

Psychoacoustic Analyzer

Waves
www.waves.com

PAZ is a real-time psychoacoustic analyzer plug-in, which shows up to 68 bands with RMS, peaks, weighting, variable integration time, and other data. It uses wavelet techniques to provide optimal graphic accuracy (see Figure 6-16). Individual bands update independently for the fastest response and resolution.

In addition to peak/RMS metering, the plug-in has a continuous-graph frequency display that shows the 52 bands that most closely resemble the ear's constant Q critical frequency bands. Optional resolution in 10 Hz steps can be shown for precise analysis below 250 Hz for a total of 68 bands. The display can be zoomed into any area of the graph. The analyzer, pan, or VU can be displayed individually or all together. The stereo position display shows energy spread in the stereo field, including antiphase

Figure 6-16
The Psycho–
acoustic
Analyzer
interface
provides the
ultimate
graphical
accuracy.

information. It can be used for mastering, trouble-
shooting, and environmental analysis.

S1 Stereo Imager

Waves
www.waves.com

S1 Stereo Imager is a set of tools for remastering stereo
mixes by enhancing and altering the stereo effect through
psychoacoustic shuffling processes. The controls are fairly
simple (see Figure 6-17). Width alters the size of the stereo
image. Rotation alters the level balance without affecting
center-channel sound. Polarity and channel swap are for
correcting stereo errors. The stereo vector display shows
the effect of stereo alterations in real time. It will process

Figure 6-17
SI Stereo
Imager's controls
are simple.

at 88.2/96 kHz with 48-bit double precision. It will dither
to 24 bits for output.

Audio Utilities

BF Essentials

Bomb Factory
www.bombfactory.com/

BF Essentials is a collection of audio utilities that are
designed to save time, solve problems, and provide user
feedback for practical solutions to everyday needs.

BF Essential Tuner is a CPU-based tuner (see Figure
6-18).

The BF Essential Clip Remover is used to repair clipped
audio recordings without a pencil tool (see Figure 6-19).

The BF Essential Meter Bridge provides RMS or peak metering, and calibrates instantly for useful viewing at any signal level (see Figure 6-20). It is more accurate than the built-in meters in some applications.

Figure 6-18
BF Essential
Tuner

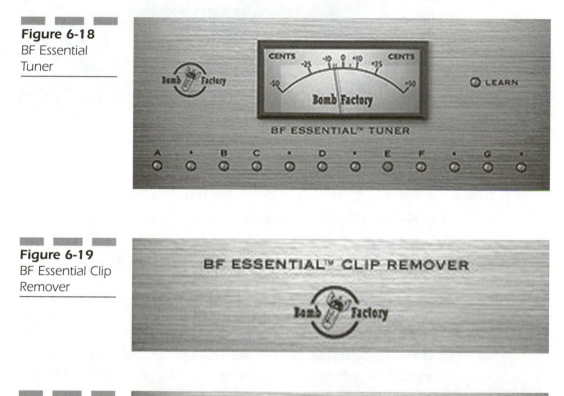

Figure 6-19
BF Essential Clip
Remover

Figure 6-20
BF Essential
Meter Bridge

Figure 6-21
The BF Essential
Noise Meter

The BF Essential Noise Meter is three meters in one (see Figure 6-21):

- Set to "A," it's an A-weighted noise meter.
- Set to "R-D," it's a Robinson-Dadson equal-loudness meter.
- Set to "None," it's a VU meter with 100 dB of visual range.

Mono and stereo versions are included.

The BF Essential Correlation Meter is used to solve tracking and mix-phase problems and troubleshoot phase co-herency. It works on stereo tracks or stereo submixes (see Figure 6-22).

SpectraFoo/SpectraFoo Complete

Metric Halo
www.mhlabs.com

SpectraFoo Complete is a multifunction audio utility that includes level metering, spectral analysis, correlation

Figure 6-22
BF Essential
Correlation
Meter

metering, triggerable waveform display, power balancing, and a variety of spectral histories and phase analysis. It can be applied to any number of input or output channels.

The plug-in has audio frequency oscilloscopes, power balance meters, level meters with physical unit calibration, audio spectrum analyzers, Spectragram spectral history meters, correlation meters, correlation history meters, frequency-sensitive phase meters, envelope history meters, and limited history meters.

It has a full-featured signal generator with 24-bit, distortion-free signal generation, up to nine simultaneous sine sweeps, pink and white noise generation, burst generation, and FFT-synchronized sine generation with direct generation to audio I/O, captures, and files.

The transfer-function measurement system allows the direct measurement of both acoustic and equipment transfer frequency and phase functions. It can be used for equipment verification and testing or to measure control rooms for acoustic and electronic corrections and many other test and measurement functions.

The system maintains a complete parameter storage and retrieval library mechanism for all instruments in the system. Modifications of parameter libraries can be saved over the originals or cloned to create new libraries. There is also a comprehensive set of instrument presets.

With SpectraFoo, you can create a complete snapshot of meter positions, visibility, configuration, channel routing, and other settings of the entire state of the system, and save it as a preset document (see Figure 6-23).

TL InTune

Trillium Lane Labs
www.tllabs.com

TL InTune is a professional tuner that emulates the features of rack-mounted digital tuners in a Pro Tools plug-in

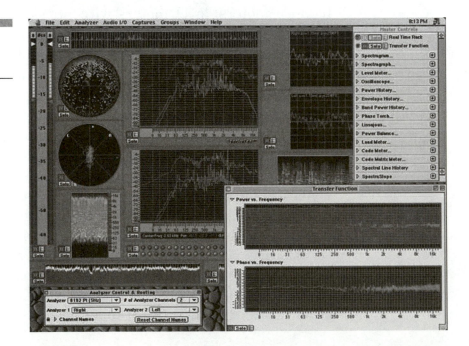

Figure 6-23
SpectraFoo
Complete

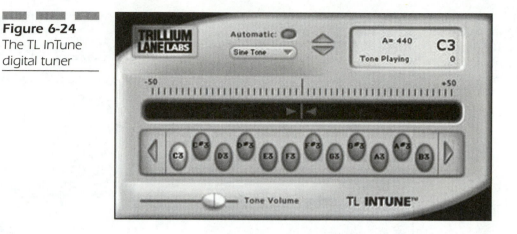

Figure 6-24
The TL InTune
digital tuner

(see Figure 6-24). TL InTune can be used for tuning musical instruments directly, in the context of a Pro Tools session. The presets include both popular and obscure tuning types for all sorts of instruments, and additional presets can be downloaded from the Trillium Lane Labs' web site. TL InTune can also act as an electronic tuning fork and produce a wide range of reference tones.

TL Metro

Trillium Lane Labs
www.tllabs.com

TL Metro is a metronome for use directly in Pro Tools recording sessions and takes the place of a click track (see Figure 6-25).

It has a range of preset metronome sounds, or samples can be imported. The output can be recorded to an audio track or played live along with Pro Tools.

The metronome can be linked directly to a Pro Tools transport. The tempo ranges from 40 to 250 *beats per minute* (BPM) and is adjustable to 1/100th of a beat

Figure 6-25
The TL Metro
digital
metronome

resolution. There is a tap tempo function for entering tempos by hand. The BPM is adjustable. There is a visual indication of accent and regular beats, as well as independent controls for accented, quarter, eighth, and sixteenth notes, and triplets.

Composition/Looping Editing Tools

The editing process could quickly become tedious if you are trying to repeat one phrase throughout a track. Looping tools allow you to automatically repeat phrases according to your design. Composition and looping tools are useful additions to Pro Tools.

ALKALI

Audio Genetics
www.gallery.co.uk/audio-genetics/index.html

ALKALI is designed to automatically fit loops to any tempo, quickly and without time-stretch artifacts. Loops can be further processed with Varispeed, Quantize, and Swing features to improve the groove. It syncs to a *musical instrument digital interface* (MIDI) beat clock coming from ProTools and works like a live groove track running alongside ProTools (see Figure 6-26).

Once you set the tempo fader, every loop will automatically fit the current tempo, without time-stretching artifacts, regardless of the loop's original tempo.

ALKALI can be used to change keys, create lo-fi versions of loops, and can use loop offsets for creating polyrhythms. ALKALI comes with a CD-ROM of REX Loops from AMG.

Figure 6-26
Audio Genetics'
ALKALI

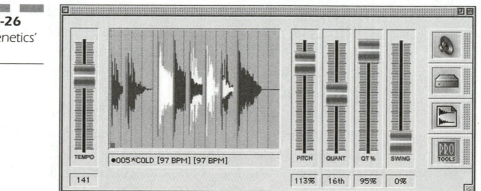

Surround Tools

If you are editing audio for a DVD, you may want to have surround-sound capabilities. The following plug-ins make surround sound a possibility.

SurroundScope

Digidesign
www.digidesign.com

SurroundScope is a multichannel metering plug-in from Digidesign. The user inserts it on a multichannel track in Pro Tools and then routes the audio for the multichannel through it so that the multichannel can be monitored by SurroundScope.

SurroundScope automatically detects the multichannel format of the Pro Tools track and places the channels in a circle around the surround display's center. Surround signals are displayed using a circular pattern. The surround display shows the user a signal's position within the surround field. A perfect circle in the center of the surround display denotes a perfectly centered surround signal. A signal that balloons toward the outer edges of the display indicates which channels have the strongest signal (see Figure 6-27).

The plug-in includes a Lissajous Phase Meter for detecting out-of-phase signals by providing visual feedback on phase coherency from perfect mono signals to completely out-of-phase signals. The phase indicator dot glows green when the signal is positively out of phase and red when the signal is negatively out of phase.

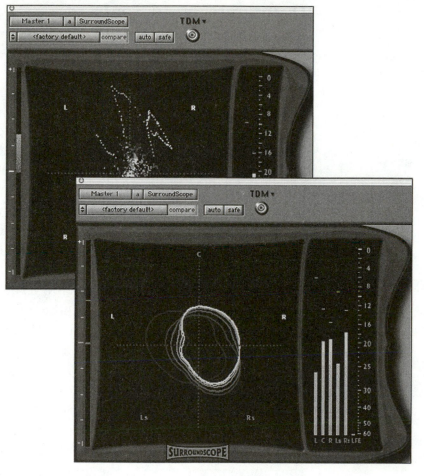

Figure 6-27
surround Scope's graphics indicate where the surround signal is strongest.

A multichannel level meter also appears as part of the plug-in. The multichannel level meter works on up to eight channels of surround mix. It provides a graphical display of the signal level for each audio channel within the multi-channel sound field.

Dolby Surround Tools

Dolby
www.digidesign.com

Dolby Surround Tools make up a TDM plug-in for Dolby surround encoding and decoding. The plug-in can be used to produce Dolby surround mixes for video, DVDs, multimedia, CDs, and video games from within Pro Tools. It is not designed for the final mixing of matrix-encoded theatrical film soundtracks. The unit is a software emulation of the encoding and decoding processes in the industry standard Dolby Model SEU4 and SDU4 hardware units (see Figure 6-28).

Dolby Surround is a phase-matrix encode/decode process that enables regular stereo program sources to carry four channels of information. Originally developed as a sound-with-picture format, the channels are configured in the LCRS format with three across the front (left, center, right) and a fourth (surround) reproduced by speakers to the sides and rear of the listening area.

There is a game mode function in the plug-in as well that can be used to preprocess mono sound effects so that they move in response to the action onscreen as the game is played. A game-mode positioner function emulates the video-game controller for confidence monitoring during production.

Figure 6-28
Dolby's
Surround Tools
plug-in

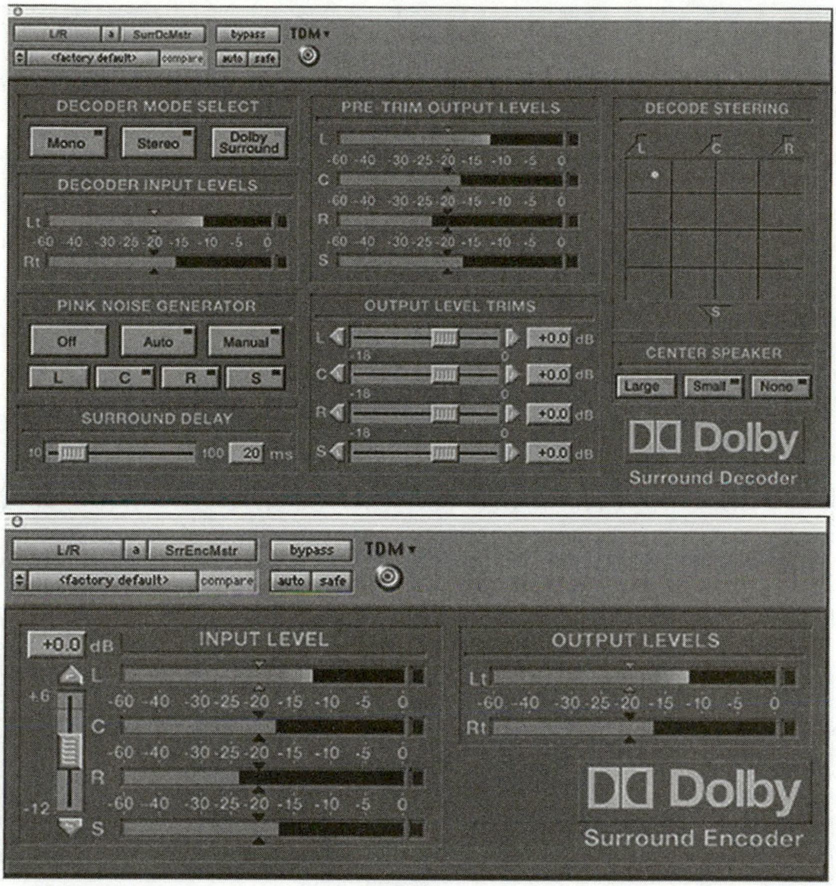

TheaterPhone HSM5.1

Lake Technology
www.lake.com.au

TheaterPhone HSM 5.1 is a Pro Tools plug-in designed to provide 5.1 surround-sound monitoring through stereo headphones (see Figure 6-29).

Figure 6-29
TheaterPhone
HSM 5.1

If you don't have a surround speaker environment to mix in, TheaterPhone can be used to create a surround mix by modeling the surround environment and delivering an emulation of the surround mix to the headphones.

There are three virtual acoustic environments. DH1 is a reference control room. DH2 is a livelier, dubbing-size room. DH3 is a larger, theater-size room. Each room has been modeled with impulse-response measurements of real acoustic environments.

TheaterPhone HSM 5.1 uses Dolby headphone technology. It is a signal-processing system that can take up to six channels of audio from any source and make it sound like it's coming from six speakers, creating a natural 5.1 soundfield and providing the senses with a perceived three-dimensional sound image, as if you were sitting in the sweet spot of a 5.1 surround speaker system.

360 Surround Toolkit

Waves
www.waves.com

Waves' 360 Surround Toolkit is a complete set of surround tools for localization, spatialization, and envelopment, with enhanced panning, reverb, and dynamics that support sample rates up to 192 kHz.

Figure 6-30
The Surround
Manager

The Surround Manager is for calibrating the studio monitor setup for standard compliant mixes (see Figure 6-30). It includes flexible bass-management features for all industry-standard surround release formats.

The Surround Reverb includes six channels of completely decorrelated reverberation with special front and rear surround control (see Figure 6-31).

The Surround Imager controls distance panning, generating early reflections and shuffling for greater low-frequency spatialization (see Figure 6-32).

The Surround Limiter is a six-channel implementation of the Waves L2 Ultramaximizer technology with brickwall, look-ahead, peak-limiter algorithms.

The Surround Compressor is a full-featured multichannel compressor that has flexible channel coupling and grouping for surround applications.

Figure 6-31
The Surround
Reverb

Figure 6-32
The Surround
Imager

The Surround Mixdown plug-in is designed to create mixes in other formats, including quad, LCR, stereo, and mono from the surround mix.

Surround Panner allows for precise 3-D placement of tracks in the surround mix.

The LFE 360° low-pass filter is an extra steep, low-pass filter for the LFE channel.

The IDR 360° Bit Requantizer provides optimal bit depth reduction.

Sampling Tools

Sampling tools provide you with libraries of samples that you can use in your audio tracks. They come in a variety of styles and can be a good addition to your Pro Tools system if you are looking for specific or eclectic sounds.

SampleTank

IK Multimedia
www.sampletank.com

Sampletank is a sample playback virtual instrument that comes with a library of high-quality mulitsampled sounds and supports samples from other libraries. It is a 16-part multitimbral virtual instrument with 128-note polyphony. It also has four FX slots per instrument and eight audio outputs.

The interface provides fast access to the sounds and sound-shaping tools, and makes shaping the sound of each instrument easy (see Figure 6-33). The tool sets the changes, based on the type of sample being processed.

Figure 6-33
Sample Tank's
interface makes
sound shaping
easy.

SampleTank comes with over 2.5GB of samples in more than 450 multisampled sound banks. The sounds are well organized and a fast, intuitive search function makes them easy to find.

The basic library includes bass, brass, drum kits, percussion, acoustic and electric guitar, piano, organ, ethnic instruments, orchestra, strings, vocals, woods, winds, analog synths, and loops. There are 28 built-in effects, including compressor, equalizer, reverb, ambience, reverb delay, delay, filter, wah-wah, chorus, AM and FM modulation, flanger, autopan, tremolo, rotary speaker, lo-fi, distortion, phonograph, and slicer.

SampleTank interfaces with Pro Tools as an RTAS plug-in.

Intakt

Native Instruments USA
www.native-instruments.com

Intakt is a rhythmic loop sampler specifically designed for playback and creative manipulation of drum and percussion loops (see Figure 6-34). It has three sample playback algorithms or modes: Beat Machine, Time Machine, and Sampler Mode.

In Beat Machine mode, it divides imported audio into individual hits. Individual settings for pitch, playback direction, pitch envelope, amp envelope, distortion, delay, and other effects can be applied.

The Time Machine time-stretches or compresses the sound in real time. It analyzes the source sound and dynamically adapts to the source sounds.

The Sampler mode plays back the source sound like a standard sampler, by linking pitch with time.

The plug-in includes a library of thousands of loops from Zero-G to East-West, covering a broad range of pop, hip-hop, and electronic styles. It has a wide range of acoustic drums, ethnic percussion, vintage jazz kits, and rhythmic guitars, as well as drum and bass, techno,

Figure 6-34
The Intakt
window

driving funk, and synthetic rhythms. It also imports WAV, AIFF, REX1 and REX2 files, along with Gigasampler, Akai, EXs, BATTERY, and KONTAKT instruments.

Kompakt

Native Instruments USA
www.native-instruments.com

Kompakt is a new sampler from Native Instruments. It utilizes the Kontakt sampling engine and format, and can also import most existing sample formats, including Gigasampler, HALion, EXS, and Akai. The plug-in comes with a library of of over 200 instruments from the East-West and Zero-G libraries.

The interface has an integrated browser with drag-and-drop functionality. The instrument parameter's controls are on the interface level and can be reached without much digging through menus. The multimode filters, envelopes, LFOs, and integrated reverb, chorus, and delay effects are easily accessible (see Figure 6-35).

Kompakt plays back samples directly from a disk, allowing for any length of sample. It has eight-part multitimbrality and 256-voice polyphony.

Kontakt

Native Instruments USA
www.native-instruments.com

Kontakt is an easy-to-use software sampler with the sound shaping and playback features of both high-end samplers and synthesizers (see Figure 6-36). It is new technology with an integrated granular-resynthesis

Figure 6-35
Kompakt's
interface

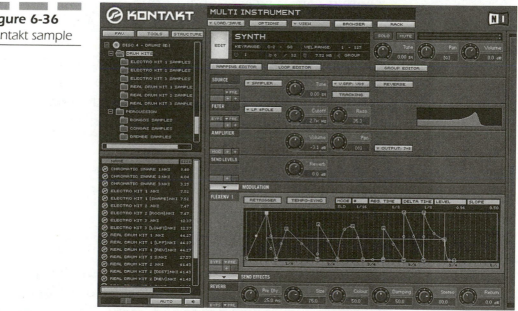

Figure 6-36
Kontakt sample

engine that makes real-time time-stretching and resynthesis possible. Pitch and time can be stretched independently.

Kontakt has integrated graphical-mapping and loop editors, and direct-from-disk sample playback. It is 16-part multitimbral with up to 256 stereo voices and 32 outputs.

The plug-in has 17 varieties of filters, from analog low-pass and high-pass, to more extreme sound-design filters. There is a broad range of insert and send effects, including EQs, wave shapers, delays, and reverbs. When effects are applied to a sample, they are stored as parts of the instrument, regardless of how many instruments are playing at once. Most parameters can be dynamically modulated by LFOs, breakpoint envelopes, step-modulators, or via an MIDI velocity or controller, and all time-based modulation can be synced to song tempo and is shown graphically.

Kontakt comes with a large library of samples and can import Akai, Battery, Gigasampler, AIFF, WAV and other formats from 8 to 32 bits.

Up to 16 instruments with their own sample maps can be assigned to individual MIDI channels to form a multi-instrument track, which can be saved and reloaded with all effect sends, filter settings, and other parameters.

MachFive

Mark of the Unicorn
www.motu.com

The MachFive is a surround-capable sampler plug-in. It is called universal because you can drag and drop samples and sample libraries in virtually all formats, including Akai S1000/3000, Akai S5000/6000, Akai MPC2000/MPC3000, Roland S7xx, EMU III/ESI/EOS, GigaSampler/GigaStudio, Emagic EXS24, Digidesign

SampleCell, Creamware Pulsar STS, and Steinberg HALion. Sample import formats include Kurzweil K2xxx, Akai MPC.snd, WAV/ACID, AIFF, SDII, and REX. A large sample library is included with the software and some instruments are sampled in surround.

Multiple instances of the sampler can be used simultaneously. Each instance has 16 parts, with corresponding MIDI channels. Complex layers and splits can be created within each of the 16 parts.

MachFive is a single-window interface and everything is done in that one window. Clicking on one of the 16 parts brings up all the parameter settings for that part. There are no deep menus or dialog boxes (see Figure 6-37).

Figure 6-37
MachFive's single-window interface

Four LFOs are available per preset, and each LFO can be routed to an assortment of destinations, including filter frequency, filter resonance, drive, pitch, pan, and amplitude. The LFOs and effects parameters sync to the sequence. There is a built-in waveform editor in both stereo and surround formats. All wave editing takes place in real time and can be performed while a sequence is playing.

The plug-in has a 32-bit internal engine, with up to 24-bit, 192 kHz sample rates. It has unlimited polyphony and very low latency. MachFive comes in most plug-in formats supported by Pro Tools, including MAS, audio units, VST, RTAS, HTDM, and Dxi.

Virtual Instruments

Virtual Instruments and Synthesizers

A huge trend today is the use of virtual instruments and software-based synthesizers. Hardware-based synthesizers have developed over decades, from modular to integrated analog to digital. Digital synthesis has developed new methods of actually creating sounds. Over the last few years, the trend has been to put all the processing into software that runs on computer hosts and only use *musical instrument digital interface* (MIDI) hardware as a triggering device for actually playing back the sounds in a musical context.

The impact of this trend cannot be overestimated. It changes the way synthesizers work at a fundamental level and brings the cost down considerably.

The following quick history will help to illustrate the origins of and differences between various synthesis concepts. This history will help in understanding the capabilities of the software equivalents compatible with the Pro Tools environment and the many evolving ways of working with sound in software.

Although MIDI was first used to get a lot of different hardware pieces to communicate and coordinate with each other for performance, the trend now is to control the software instruments within the computer using MIDI controllers. The various hardware synthesizers are beginning to be replaced by computers controlled by MIDI keyboards (and other control surfaces) that have no onboard sound capabilities. As this trend continues, the development of software-based instruments and multi-instrument environments or studios is going beyond the emulation of classic hardware-based synthesizers, and creating entirely new concepts in music composition and sound design using digital technology to combine various synthesis techniques into ever more powerful combinations.

Most of the current virtual-instrument technology is based on some form of past hardware technology. The following section is a brief description of the types of synthesizers that have shaped the thinking up until now. However, recent indications show that the flexibility that is inherent in software is beginning to blur the lines between the different approaches to synthesis. Some of the newer products, in particular, are combining functions and extending capabilities in ways that could never have been done in hardware formats before.

Modular Synthesizers

The earliest synthesizers were actually combinations of individual hardware boxes or *modules* that each had a specific function. There were boxes with signal generators, such as *voltage-controlled oscillators* (VCO) and noise generators, and other boxes with signal processors, such as amplifiers and filters. The various boxes were connected with patch cords in combinations to create sounds. A unique sound created by a combination of boxes patched together was called a *patch*, a term that is still prevalent today in describing a particular programmed sound.

Of the early modular synthesizers, the Moog became the most famous. *Switched-On Bach*—a popular release by Wendy Carlos—introduced the modular synthesizer to the general public. The design features of the Moog became the industry standard and the basis for subsequent industry innovation.

Early modular synthesizers were usually very big and somewhat unwieldy, making them impractical for transport and performance, but they offered tremendous freedom in creating new sounds. The strengths and limitations of the modular synthesizers led to the development of prewired synthesizers, which took the most common modules and routing patterns of a modular synthesizer and added various control options for selecting different patches quickly and efficiently. This innovation made synthesizers more practical for musicians and led hardware synthesizers to the mass market.

The virtual modular synthesizer was the next evolution in synthesized instrumentation. Using virtual patch cords and modules, the entire concept of modular synthesizers is now emulated in a software environment. The model and process presented to the user is based on connecting the virtual modules together to create unique patches representing custom configurations and settings of the synth modules.

Analog Synthesizers

Most simply stated, analog synthesizers are hardware synthesizers with analog-based circuitry. They have oscillators, filters, and other components controlled by voltage, and are known for having a warm, pleasing sound, compared to most digital synthesizers. Analog synthesizers work like the early modular synthesizers, except they are integrated and use internal switching rather than patching to connect the modules.

The Minimoog was one of the first analog synths. It had discrete circuitry with transistors, resistors, and other electronic components, and had no integrated circuits. Later synthesizers began using more compact analog technologies, including linear integrated circuits. As digital technology developed, more digital circuitry was used in analog synthesizers, initially for memory functions and increasingly for processing functions. Instead of VCOs, some systems introduced *digitally controlled oscillators* (DCOs). These systems are sometimes referred to as "analog hybrids."

Digital Synthesizers

Basically, digital synthesizers are synthesizers that use digital methods to generate sounds. The technology was first introduced in the 1950s. They remained somewhat impractical and very expensive for a long time. The New England Digital Synclavier workstation and the Fairlight sampler were the first practical units that achieved widespread use, but they were limited by the fact that they cost a small fortune. When the Yamaha DX7 came out in 1983 for about $2,000, digital synthesizers became popular. The DX7 was a mono synth; later entries in the DX line became polyphonic.

Except for the actual production of sound, digital synthesizers are similar to their analog counterparts in design, concept, and workflow. Rather than a VCO, they use digital samples or computer-generated waves, and process them through various digital algorithms.

The Yamaha DX7 used digital methods to do a type of synthesis called *frequency modulation* (FM). The Roland D-50 used short *pulse code modulation* (PCM) samples, or digitally generated waveforms, to generate sounds.

Sample-Playback Synthesizers

Sample-playback synthesizers contain hard-coded samples stored onboard the ROM of the synthesizer. These hard-coded samples are used as a sound source, and the signal is then processed through various filters, envelopes, and other digital circuits. They are also referred to as *wavetable* synthesizers. Sample-playback synthesizers tend to be highly polyphonic, realistic sounding, and relatively easy to use.

In the world of Pro Tools and other digital audio workstations, there are many interesting virtual versions of digital synthesizers.

Samplers

A sampler is a device that can take an external sound and make a digital sample of it, process it, and play it back. There are a wide variety of hardware-based samplers on the market and a growing number of software-based packages include a substantial sampling capability in conjunction with numerous other features.

Software Synthesizers and Physical Modeling

The idea behind modeling is to use software code to produce the sound, instead of using hardware methods. This is an exploding area. There are now several hardware-based synthesizers that have modeling capabilities. For most of them, there is a software version as well. A growing number of software products perform many functions based on physical modeling, including entire conceptual studios. For instance, Propellerheads' Reason uses the concept of a virtual rack, into which the user can insert and connect virtual instruments and outboard components as needed to build a custom studio completely inside the computer.

TDM Synthesizers

Synthesizers based on *time division multiplexing* (TDM) are powered by Pro Tools Mix or HD systems *digital signal processor* (DSP) chips, instead of the computer host's *central processing unit* (CPU) power. DSP-based synths have the same advantages as TDM effects plug-ins in that they provide a solid level of real-time performance and don't drain the host machine's power.

The currently available TDM synthesizers include Access Virus, DUY SynthSpider, and McDSP Synthesizer One. Waldorf Q TDM is expected soon.

Virus Indigo

Access Music
www.access-music.de

The Virus TDM plug-in was the first plug-in to utilize TDM architecture, using the TDM DSP chips for processing, instead of host processors. The successor is the Virus Indigo plug-in. It's based on the current Access Indigo 2 hardware synth. It uses analog modeling to create timbres that closely resemble classic analog synthesizers.

You can use a maximum of eight plug-in instances per session, with as many as 20 voices per DSP chip. A Pro Tools HD system running at 44.1 or 48 kHz will support up to 160 voices. At 96 kHz, it can handle 60 voices on HD. Expanded Pro Tools mix systems can handle up to 16 voices per DSP for a total of 128 potential voices per session.

There are 1,000-plus preset patches. The first page in the Indigo hierarchy, called the Easy Page, provides access to some of the patch's main controls (see Figure 7-1). There are other, deeper pages for all kinds of fine control.

There are two main effects pages. FX-1 includes analog boost, chorus, phaser, distortion, a ring modulator, and a detailed delay/reverb section. FX-2 has global controls, a vocoder section, an arpeggiator, an envelope follower, and the external input selector.

Virus patches are divided into 17 library preset categories: Init Sounds, Acid, Arpeggiators, Bass, Classic,

Figure 7-1
The Easy Page
of Virus Indigo

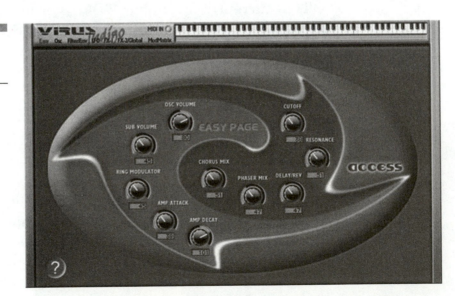

Decay, Drums, Effects, Input, Leads, Organ, Pads, Percussion, Pianos, Pluck, Virus Indigo, and Vocoder. The quality of the patches is generally very good. You can, of course, start with any preset, and then program and save your own patches.

The integration with Pro Tools is very tight. Automation is available in two modes: normal Pro Tools automation or MIDI automation. With the Pro Tools automation, you choose a parameter, put it in Auto-Write mode, and the program tracks your moves. With MIDI automation, you can use a MIDI-compatible control surface to write MIDI automation, and you can map a MIDI control surface with a selection of parameters.

SynthSpider

DUY Research
www.synthspider.com

SynthSpider is a virtual modular synthesizer for Pro Tools TDM and HD systems. It allows you to build your own synthesizers from virtual modules. The concept is to link a custom selection of the 40 different modules to build custom synths for a wide range of sounds and purposes.

The modules include the following:

- Pitch tracker
- Ramp generator
- Shaper
- Envelope follower
- Envelope generator
- Dynamic wave generator

- Sequencer
- Mixer
- One sample delay
- Wave scanner
- Karplus-Strong
- Chorus
- MIDI note module
- MIDI velocity module
- MIDI gate module
- MIDI control module
- Oscillator
- Triangle oscillator
- Square oscillator
- Noise generator
- Sample and hold
- One-pole low-pass filter
- One-pole high-pass filter
- Two-pole low-pass filter
- Two-pole high-pass filter
- Band-reject filter
- Band-pass filter
- Slider
- Plasma meter
- Absolute value
- Text label
- Scale

- Signal oscilloscope
- Shift right/shift left
- Numeric value display
- Invert signal
- Addition, subtraction, and multiplication

Ultimately, the range of sounds is huge because the various synthesizers the user creates have completely differently structures.

This synthesizer is usually monophonic. It's possible to play SynthSpider polyphonically, but you need to insert as many instances of the plug-in as you require voices for polyphony and each one needs to be on a different track in Pro Tools. You can then set up instances of SynthSpider as slaves to a master instance, and any patch changes or edits you make with the master instance are automatically copied to its slaves.

SynthSpider also has several sets of presets with virtual analog sounds, FM and physical modeling presets, territory, effects patches, vocoders, and emulations of acoustic and electric instruments (see Figure 7-2).

Synthesizer One

McDSP
www.mcdsp.com

Synthesizer One is a two-oscillator wavetable synthesizer for Pro Tools TDM and HD systems. It is not multitimbral, but polyphony can be obtained by using multiple instances. A Mix-SRAM chip supports up to eight voices and an HD chip can handle 10 voices.

Figure 7-2
The SynthSpider
interface

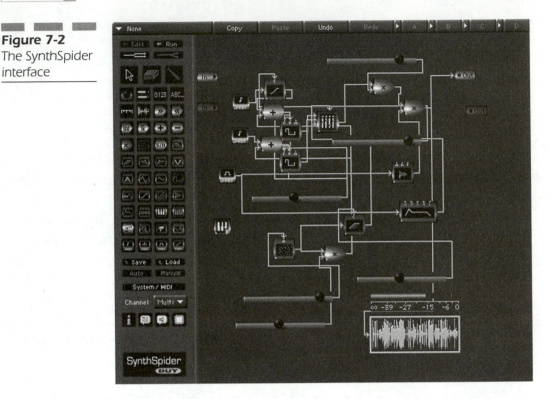

The synthesizer has two wavetable oscillators; three wave sources with independent waveform, level, phase, octave, PW, and PWM control; FM and FM velocity controls for each oscillator; a separate ring oscillator with selectable inputs; a separate noise source; an analog-style oscillator; three *low-frequency oscillators* (LFOs); and various filters, delays, and other effects. It has a modular design that permits flexible routing between the various features, including external audio (see Figure 7-3).

Synthesizer One comes with a collection of presets and the ability to define user presets. It is capable of creating a broad range of analog synthesizer sounds, particularly vintage lead, bass, and percussion sounds.

Figure 7-3
The modular-
style Synthesizer
One

Figure 7-3
The modular-
style Synthesizer
One

Waldorf Q TDM

Waldorf
www.digidesign.com
www.waldorf-music.com

The Waldorf Q TDM is a synthesizer plug-in that is an emulation of the Waldorf Q synthesizer, a powerful hardware synth. This plug-in is not yet shipping as of this writing and limited information is available, but all indications are that this should be a high-end synth module capable of analog-type sounds and FM synthesis (see Figure 7-4).

RTAS and HTDM Synthesizers

RTAS, HTDM, and native synthesizers are powered by the host computer's RAM. There is a wide range of these instruments available.

Figure 7-4
The Waldorf Q
TDM plug-in

Kantos

Antares Audio Technologies
www.antarestech.com

Kantos is a software-based synthesizer that, unlike conventional MIDI synthesizers, is controlled by audio. Kantos analyzes incoming audio from any monophonic source, including voice or instruments, and extracts pitch, dynamics, harmonic content, and format characteristics in real time. It then uses the information to control its sound engine (see Figure 7-5).

The sound engine has two wavetable oscillators, pitch constraint and quantization control, a noise source, three resonant multimode filters, two chorus generators, two envelope generators, two LFOs, a modulation matrix, a gate generator, a noise gate, delay line, and multiple mixers. It has a unique feature called a *timbral articulator* that takes the harmonic content and format information

Figure 7-5
The Kantos
interface

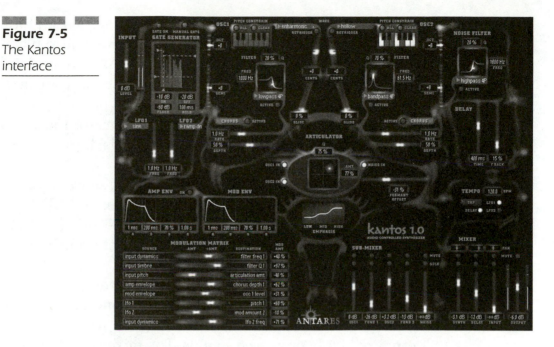

from the input signal and dynamically applies it to the synthesized signal.

The basic concept is that you can input a dry mono line from any instrument and output a fully developed synthesizer sound without ever accessing any MIDI. The synth engine actually derives its sound from the incoming audio signal's harmonic and format content. You can then apply a full range of synth controls on top of that audio line.

To use the plug-in, inset Kantos on an audio channel and send it some signal. Then, set the input-level control, followed by the gate generator, and a combination noise gate and trigger generator. Threshold, hold, and note on and off values trigger the oscillators. The gate and trigger parameters can be adjusted in the real-time waveform display.

The main tone generators are two wavetable oscillators. Kantos comes with a selection of wavetables, and you can create additional custom wavetables as .AIFF and .WAV audio files. Each oscillator has pitch-constrain control, a filter, and a chorus effect.

Both oscillators and one noise generator are routed to the articulator, where the incoming signals' format information and harmonic content are applied.

The modulation matrix routes the input sources, which include dynamics, timbre, pitch, envelopes, and LFOs, to the destinations that include filters, fine-tuning, articulation, format offset, chorus, delay, and the modulation matrix's own modulation amount settings.

Then the submixer mixes the oscillator outputs with a sine wave based on the input signal's fundamental frequency, which serves to anchor the pitch and frequency range of highly processed sounds. The mixer sets the overall level and pan in relation to a delay and the dry input signal.

Lounge Lizard

Applied Acoustics Systems
www.applied-acoustics.com

The Lounge Lizard EP-1 electric piano is a fairly simple instrument. A note played on the keyboard activates a hammer that hits a fork. Then a magnetic coil pickup, similar to an electric guitar pickup, converts the sound to an audio signal and sends it to an amplifier stage and output.

The concept of the Lounge Lizard is to emulate the exact physical process in software, providing controls for the user to tweak each stage. The idea is that this technique is the best way to produce sounds that sound like the real thing, since certain physical peculiarities in the hardware instrument will only be created by mimicking (or physically modeling) the entire chain of events that produce the original sound. For instance, striking a fork or tine on a Fender Rhodes piano will produce a different sound when it is still in motion from a previous strike than when it is at rest. In Lounge Lizard, software modules correspond to the major physical parts of the hardware instrument, including the mallet, the fork, the pickup, the amp, and its effects. When it receives a MIDI command to play a note, the plug-in generates a mallet strike that passes through the modules in the right order to recreate the physical process and its unique characteristics. The user can tweak the components in the same way one might use a screwdriver to make adjustments to the physical unit's components.

The mallet module has stiffness parameters that allow you to change the mallet's material, stiffness, and force parameters that determine how hard the mallet hits the tone bar. The noise parameters and keyboard scaling further refine the nature of the mallet hit.

The fork module has two parts—the tone bar and the tine. The tine is the part that is struck by the mallet. It controls pitch, volume, and decay. The tone bar resonates passively when the tine is struck and its pitch, volume, and decay can be adjusted separately.

The pickup module is the next step. Its symmetry parameter adjusts the virtual pickup's vertical position in relation to the tine. In a real electric piano, moving the pickup in this manner changes the amount of overtone and alters the overall timbre. Similarly, the distance parameter, as

the name implies, adjusts how far the pickup is from the fork, which changes the tone as well as the volume. The amp controls adjust the signal level for pre- and post-pickup.

The Lounge Lizard has several effects modules, including wah, phaser, and tremolo, and a stereo delay modeled after classic analog delay boxes. There is a collection of good presets, but the unit is really designed for creating custom piano sounds (see Figure 7-6).

Tassman

Applied Acoustics Systems
www.applied-acoustics.com

The Tassman Sound Synthesis Studio is a modular-instrument and sound-design environment based on physical modeling. Physical modeling is a very different approach to virtual instruments than sampling.

In physical modeling, the laws of physics are used to reproduce the behavior of an object in a computer-based emulation. Mathematical equations describing an object's functions are created and then solved in real time according to the performance controls that the program receives via MIDI.

The approach works by reproducing the way an object creates sound by reproducing the behavior of the real object rather than trying to reproduce the sound signal itself with wavetables, additive synthesis, samples, or other methods.

Tassman comes with a collection of 50 prepatched instruments, and over 1,000 presets, including classic analog and FM instrument emulations, acoustic instruments, an electric piano, tone-wheel and pipe-based organs, and many others. The browser organizes all of the modules, subpatches, instruments, and presets into hierarchical groups.

The main modules are the Player, the Builder and the Browser, which are all integrated into a single window.

In the Player view, you can tweak the existing presets to create custom sounds and save them for instant recall. In the Builder, you can customize the way individual modules are connected within a synth, or create instruments from scratch. The plug-in comes with a collection of preconfigured groups of commonly used generators, filters, and effects called *subpatches*, and you can create and save custom subpatches. The Browser displays all of your patches in a logical system of progressively more detailed menus (see Figure 7-7).

Figure 7-7
The Tassman
browser uses a
series of
progressive
interfaces.

Unity Session

Bitheadz
www.bitheadz.com

Unity Session is a combination of two previous products—
the Retro AS-1 software synthesizer and the Unity DS-1
software sampler. It adds a new physical modeling technology, and a new system-level processing component
called Unity Server that automatically launches any time
a service is required by either Unity Session or another
program. Everything else in Unity Session is a plug-in to
Unity Server.

Unity Session supports up to 24-bit, 96 kHz audio and
has up to three stereo oscillators per synth, which elimi-

nates many syncing problems. You can apply and read splice points in loops, change the tempo, and synchronize to a MIDI clock without changing pitch.

The plug-in can stream samples from RAM or directly from a disk. Unity Server can also record its own output back to disk. You can trigger multiple channels live from a sequencer and sample the result as a form of resampling.

The architecture is based completely on the concept of plug-ins. The original Unity DS-1 and Retro AS-1 along with the new sound-generation sources, such as physical-modeling modules, are all plug-ins to the main program. Real-time translators for nonnative sample formats are also plug-ins. It can read Giga, Akai, Roland, SampleCell, Sound Designer, AIFF, WAV, SoundFont 2.0, DLS, CD audio, and other formats. You can use the different types of sound sources and effects simultaneously. The Unity control panel has nine tabs that divide the basic controls into panes dedicated to configuration, bank locations, audio, MIDI, plug-ins, and so on.

There are four separate applications in Unity Session: Unity Mixer, Unity Editor, Unity Keyboard, and Unity Player.

Unity Mixer is for the setup of MIDI channels, including mixing sound module plug-ins (see Figure 7-8).

Each channel has two inline MIDI effects, two in-line audio effects, two audio-effects sends, transposition, tuning, pan, and volume. The master section also has two inline MIDI effects, two master audio-effect sends, and two master inline audio effects.

Unity Editor is where you tweak the sounds, including synthesis, sampling, and modeling. In this section, you can chop up samples, add crossfades between as many as four

Figure 7-8
The Unity Mixer

layers of samples on a single oscillator, and mix samples together to create a new one. You can also define tuning, controller maps, and custom audio routing. You can also resample, shift pitch, stretch time, loop syncs to an external clock, and edit MP3 files.

The Unity Player application is a simplified version of Unity Mixer that is used for performance. The fourth application, Unity Keyboard, is for testing setups from the keyboard prior to performance.

Unity Session comes with several CDs of sounds, including the "Best of Retro" and "Best of Unity" banks and three BitHeadz library titles, "Black & Whites," "Pop Drums," and "Orchestral Strings." Also included is the Osmosis application, which can convert Roland S760/770 or Akai S1000/3000 sample CDs into Unity 3.0, SampleCell, or AIFF format.

Bruno/Reso

Digidesign
www.digidesign.com

Bruno/Reso is a set of two synthesizer plug-ins that have similar user interfaces and use the same sound generation algorithm. They are both controlled by an incoming audio signal. They take the audio track and run it through a variety of synthesizer modules and the resulting sound can then be played via a MIDI controller.

Bruno and Reso handle the audio source track differently. Bruno slices pieces of the source track and cross-fades them together, continually blending the sound with natural sounding results (see Figure 7-9). In its Timbre

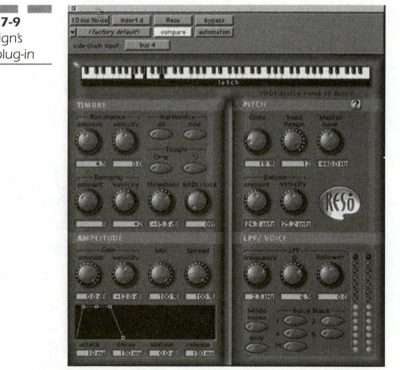

Figure 7-9
Digidesign's
Bruno plug-in

section, Bruno has a crossfade dial for adjusting the rate at which slices are culled from the source track and faded together. Slices can be butted rather than crossfaded, which creates a choppier, more rhythmic, effect.

Reso uses the audio source to drive the synthesis of new harmonic overtones, resulting in a wide range of diverse sounds (see Figure 7-10). Its controls include resonance and high-frequency damping for controlling the resonator's edge.

Figure 7-10
The Reso
interface

Emagic HTDM Software Instruments

Emagic
www.emagic.de

Emagic's line of software implements virtual synthesizers, very specific emulations of classic analog sounds, a sampler/synthesizer, and a sample playback device (see Figure 7-11).

There is a host TDM enabler in the Emagic software that makes it possible to use the Emagic software instruments as HTDM plug-ins with Pro Tools.

The ES1 is a virtual-analog synthesizer for the Logic series. Depending on available CPU power, up to 64 ES1 instances, with up to 16 voices each, can be used simultaneously.

It comes with a library including pad, bass, percussion, effect, and instrument patches.

Figure 7-11
One of Emagic's
HTDM software
instruments

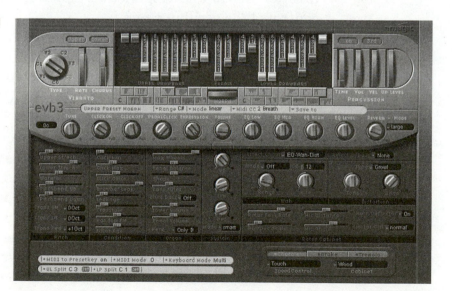

The ES1 has a main oscillator and suboscillator, each with a set of waveforms, including triangle, sawtooth, and variable pulse. The suboscillator operates one or two octaves below the main oscillator. It can also be used as a noise source or an input for an external audio signal, allowing audio tracks to be processed with rhythmically changing filter movements.

There is also a low-pass filter with an adjustable overdrive included in the bundle, as well as a fat, low-end filter and three selectable filter slopes. Other features include an LFO and a modulation matrix that allows the modulation of pitch, pulse width, mix, cutoff, resonance, volume, filter FM, and LFO amplitude. The LFO can be synchronized with the tempo in single or multiple bar/beat steps. An audio track can be used as a modulation source. The modulation envelope can control the LFO for effects.

The ES2 is a software synthesizer that supports up to 64 voices per instance. It has three oscillators per voice with analog and digital waveforms, noise FM, synchronization, PWM and ring modulation, and dynamic vector control of the oscillator mix.

Each voice has one multimode filter with overdrive and one low-pass filter that can be connected in serial or parallel. There is also a mono LFO and a polyphonic LFO.

EVP73 is an emulation of the Fender Rhodes electric piano with real-time tone generation and full polyphony over a range of 73 notes, with 73 voices. There are controls for decay, release time, and stereo spread; controls to emphasize the bell or damper noises; and an extensive tremelo. The EVP88 is similar, except it supports 88 notes and voices and a wider range of electric piano emulations, including Rhodes, Wurlitzer, and Hohner pianos. The EVD6 is an emulation of the Hohner Clavinet and the EVB# is an emulation of the Hammond B3 organ.

The EXS24 mk II is a sampler that also has a collection of synthesis tools for sound modification. It supports resolution up to 24 bit/192 kHz and can import samples in AIFF, WAV, and SDII formats. The EXSP24 is a software sample player.

impOSCar

Gmedia Music
www.gmediamusic.com

The impOSCar is an emulation of the classic British synth the OSCar. It is capable of loading original sounds via sysex and tape (via WAV files). It is fully polyphonic and has a range of keyboard velocity responses and additional filter modes (see Figure 7-12).

M-Tron

Gmedia Music
www.gmediamusic.com

The M-Tron is a software simulation of the classic Mellotron (see Figure 7-13).

Figure 7-12
The digital version of OSCar

Figure 7-13
The M-Tron
digital keyboard

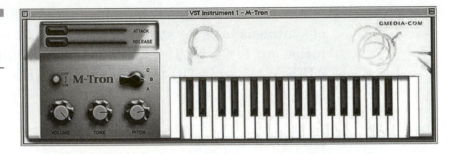

It has 28 classic tape banks that emulate the original instruments sounds. On the original Mellotron, when a key was pressed, a tape head moved onto a piece of tape to play the note or sound. For M-Tron, each of the original tape sounds was sampled from the original and then they are played back when a key is played. The original Mellotron tape loops were eight seconds in length and looped seamlessly if a note was sustained beyond eight seconds. M-tron emulates the process digitally. It has both on-screen and MIDI control over volume, tone, pitch, amplitude envelope attack, and release.

Oddity

Gmedia Music
www.gmediamusic.com

The Oddity is an emulation of the classic ARP Odyssey. The name seems to come from the recognition that the ARP Odyssey had some very idiosyncratic characteristics, which every effort has been made to capture and reproduce in this software synth. Extensive sampling of a wide range of original Odyssey's output sounds at various settings provide the basis, and a detailed analysis of the control paths provides the basis for the model and code that form the Oddity.

Figure 7-14
Oddity's
interface

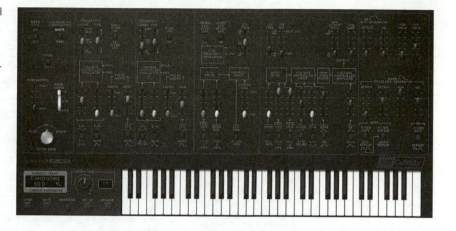

The Oddity has two syncable oscillators that are fully tunable across a six-octave range, with sawtooth, sine, square, and variable pulse-width waveforms. Its other features include white and pink noise generators, syncable LFO, a ring modulator, low-pass and high-pass filters, sample and hold, portamento, two envelope generators, and many modulation routing options. It has monophonic or duophonic modes. There are five banks of 64 presets.

The GUI has a glissandi keyboard and all parameters can be mapped to external MIDI control (see Figure 7-14).

Reaktor 4/Reaktor Session

Native Instruments USA
www.native-instruments.com

Reaktor 4 is a modular sound studio that includes a wide range of synthesis, sampling, and effects that musicians and engineers can use to design and build their own instruments, samplers, effects, and sound design tools (see Figure 7-15). It comes with a library of dozens of instruments and effects, and there is an online user library.

Figure 7-15
The Peaktor
window

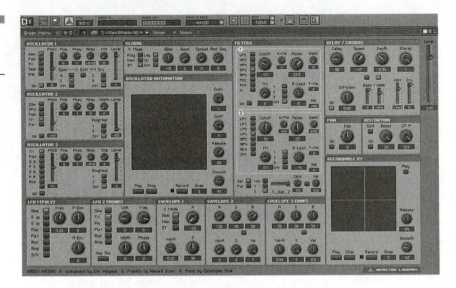

Reaktor is the core of the Native Instruments line, with its NSP technology, from which other instruments, including FM7, Pro 53, and others, are built.

The user can select from over 100 modules and macros, including oscillators, filters, samplers, sequencers, and effects, and connect them in flexible possibilities. The effects include multiband compressors, spatializers, surround reverbs, and a vocoder. Existing instruments can be combined and enhanced, or cannibalized for parts to create other instruments. The control surface of new instruments can also be customized, including the size and layout of the knobs and displays. Images and graphics can even be imported for use in the interface.

Several instruments can vary speed and pitch independent of each other, while others completely freeze the sound in time. Samples can be easily loaded via drag and drop into a comfortable, graphical sample mapper. There are four sequenced synths, as well as analog-modeling, FM, and wave-table additive synths. Just about every tool you could think of is included.

Reaktor Session is a music production tool that offers access to all of the Reaktor instruments, but excludes the instrument-building capabilities of Reaktor 4. Both are fully MIDI controllable and have the capability of working at sampling rates of up to 192 kHz.

Absynth 2

Native Instruments USA
www.native-instruments.com

Absynth is a software synthesizer with a semimodular architecture and clean interface. It combines subtractive, FM, AM, ring modulation, and wave-shaping synthesis techniques. It has six oscillators, four filters, three ring modulators, one wave shaper, and one delay processor per voice. In the waveform editor, the spectrum mode and the waves themselves can be used to trim individual harmonics. The waves can be mixed, modulated, filtered, or fractalized and then used by oscillators or LFOs and as wave-shaping distortion functions. The plug-in allows you to hear the results continuously while you edit.

There are more than a dozen envelopes with up to 68 breakpoints each. Envelopes can loop or trigger repeatedly to create cyclic rhythms and constantly evolving ambient textures. The Editor window displays multiple envelopes simultaneously, so you can precisely synchronize the evolution of different layers of sound in time against a tempo grid.

The parameters can be modulated by MIDI continuous controllers, aftertouch, velocity, and note number. Multiple instances can be used to create large ensembles. Absynth comes with hundreds of patches for a wide range of preset possibilities (see Figure 7-16).

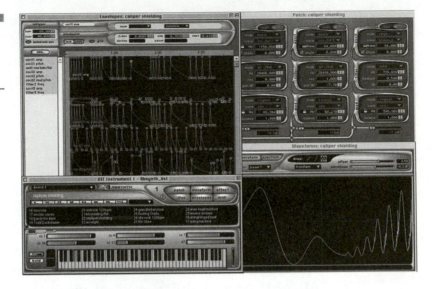

B4

Native Instruments USA
www.native-instruments.com

The B4 is a modeled emulation of the Hammond B3 organ, based on an analysis of the electro-mechanical sound generation and tube circuitry of organ and speaker cabinet combinations (see Figure 7-17). The plug-in is based on the same technology as the Reaktor modular system.

There are extensive subtle details in the model, such as harmonic foldback, drawbar crosstalk, and loudness robbing. You can apply combinations of additional effects, including scanner vibrato, tube overdrive, and rotating speaker effects.

Completely new sounds can be added by replacing the virtual tone wheels. By changing the tone wheels you can turn the B4 into a Vox Continental, Farsifa, or an Indian

Figure 7-17
The B4, a plug-
in version of the
Hammond B3

Harmonium. The tone wheels can also be used to alter the master tuning and to age the sound in six different levels, from "well matured" to "far beyond repair."

FM7

Native Instruments USA
www.native-instruments.com

The FM7 is an FM synthesizer that emulates the Yamaha DX line of hardware sythesizers. It can read all programs from the original DX7, DX7-II, DX11, TX81Z, DX21, DX27, DX100, and TX802 and reproduce the sounds of these machines exactly.

The library includes 256 presets with both standard and new sounds. In addition to the standard presets, there is a full set of programming tools. There are dedicated analog-style controllers, distortion and filter operators, extensive modulation capabilities, a comprehensive effects section with stereo chorus, flanging and delay effects, and audio input (see Figure 7-18).

Figure 7-18
NI's FM7

Pro-53

Native Instruments USA
www.native-instruments.com

The Pro-53 is an emulation of the classic Sequential Circuit Prophet 5 analog synthesizer (see Figure 7-19).

It has a new oscillator technology with a warm sound, a high-pass filter mode and a filter envelope that may be inverted. It uses the same *native signal processing* (NSP) as Reaktor and FM7. There is no fixed limit to the number of voices.

The Pro-53 can read and accurately reproduce all the existing sound libraries made for the original Prophet 5.

It comes with a library of 512 classic and modern analog sounds, and 512 preset memory slots are available for user-created presets. There is a built-in effects section with multiecho, chorus, and flanging effects. The included

Figure 7-19
The Pro-53

preset sounds make extensive use of the effects. The plug-in also has an audio input for processing external audio signals through its filters and effects. Multiple instances can be run in parallel <parallel mode? parallel tracks?>.

Phrazer

BitHeadz
www.bitheadz.com

Phrazer is a loop-based composition and arranging studio for the Mac (see Figure 7-20). It can connect with Pro Tools 6.1 via ReWire for audio output.

There are over 900MB of loops included and several CDs of additional sounds available. There are pitch- and tempo-matching algorithms for matching up different audio loops. It performs multitrack audio editing and playback with unlimited tracks available. Two send effects are available per track, including phrase, flange, chorus, degrade, distortion, filter, dynamic filter, compressor, delay with MIDI sync, shelf/parametric EQ, and reverb.

Figure 7-20
The Phrazer
studio interface

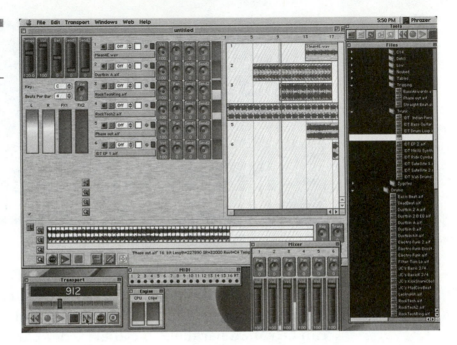

Phrazer can import MP3s, ACID I and II files, Unity DS-1 samples, AIFF, Sound Designer II, WAV files, and CD audio. It can sync to MIDI clock for integration with studio hardware and software.

Reason

Propeller Heads
www.propellerheads.se/

Reason is a truly unique music composition studio and expandable music workstation. It uses a visual metaphor of an empty equipment rack that the user then fills with various modules, such as mixers, synthesizers, and drum machines, to create a custom environment. Reason is expandable up to the practical limit of the computer's ability to process. The audio output connects to Pro Tools via ReWire, an RTAS plug-in also from Propellerheads.

Each unit in Reason's virtual rack is edited from its own onscreen front panel (see Figure 7-21). Each module has an array of sliders, knobs, buttons and functions that work in real time. All filter adjustments, pitch bending, gain riding, or panning, along with other front panel functions, can be recorded and automated in the integrated sequencer.

Changing the configuration of Reason is an easy process. By hitting the Tab key, you can flip the rack

Figure 7-21
One of Reason's onscreen front panels

around. When the back of the rack is displayed, all the cables and connector jacks are visible. To change the audio routing, just click and drag a connector to a new location.

Reason is a very original approach to creating music in an all-software environment.

Atmosphere

Spectrasonics
www.ilio.com/spectrasonics

The Atmosphere Dream Synth is a large-scale modular-synthesis environment (see Figure 7-22). It is a programmable synth with a very large high-resolution core wavetable dedicated to sound design elements, as opposed to stock instrument emulations.

The plug-in has a large core library of all kinds of sounds that were created using over a hundred sound design devices' various synthesis methods, including granular, additive, wave table, wave scanning, grain table, neural processing, vintage analog, vector, virtual analog, FM, and acoustic sources.

Figure 7-22
The Atmosphere
interface

The 1,000 or so patches can be layered and each layer can be independently tweaked with its own multimode resonant filters, four LFOs, three envelopes, and matrix modulation. There is also a master filter for quick tone shaping.

Stylus

Spectrasonics
www.ilio.com/spectrasonics

Stylus is a software instrument that integrates a core library of thousands of groove elements, loops, and samples, with a powerful user interface for creating your own rhythm loops and grooves (see Figure 7-23).

The core library has break-beat remix loops that are oriented toward variations on modern dance, R&B, and hip-hop.

Pitch, tempo, feel, and patterns of grooves can be changed independently. Grooves can be programmed from scratch using thousands of drum samples, including 1,000 kicks, 1,000 snares, and over 500 hi-hats. You can remix loops in real time at any tempo by selecting a patch and

Figure 7-23
Stylus allows you to create loops and grooves.

playing a keyboard. There are also over 1,000 turntable FX and DJ tricks included, and a live percussion loop section with congas, bongos, djembes, shakers, triangles, agogos, and tambourines, which can be mixed into any groove separately. Each sample and slice has its own adjustable synth parameters. There is also a built-in groove auditioning system and a built-in patch management system.

Stylus is fully programmable and has sample-accurate timing, with total sequence recall. There are multimode resonant filters for each sample, as well as master filters. There are three envelopes for pitch, filter and amplitude, and matrix-style modulation routing, with two LFOs.

There is independent parameter control of each sample and full groove control automation.

Trilogy

Spectrasonics
www.ilio.com/spectrasonics

Trilogy is a programmable, sample-based sound module for bass sounds (see Figure 7-24). It has three distinct sets

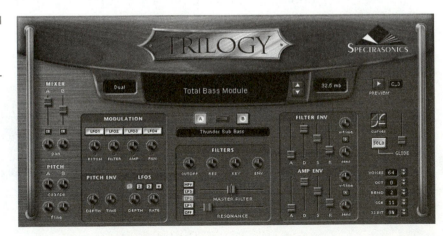

Figure 7-24
The Trilogy Total
Bass module

of sounds and tools for acoustic, electric, and synthesizer bass sounds. Each patch has two layers that you can tweak individually and mix and match with any of the layers in the core library to create combination sounds. Each layer has its own multimode resonant filters, four LFOs, three envelopes, an adjustable sample start, and matrix modulation. There's also an additional master filter for quick tone shaping.

The acoustic upright bass interface has separate control for two signals. The Neuman U-47 tube microphone signal was sampled using the vintage microphone and the direct pickup signal was sampled through a vintage Neve 1083 console. The samples are chromatic, with multidynamic velocity switching. A sampled acoustic Martin Dreadnought bass guitar is also available in the acoustic section. A technology called *true staccato* enables realistic phrasing of closely repeated notes. There is also a finger noise layer, which triggers noise samples from a library of hundreds of tiny fret noises, subtle string scrapes, squeeks, taps, and mutes, which add a realistic element of human imperfection to performance.

Trilogy has a large selection of four-, five- and six-string electric basses performed in fingered, picked, muted, slap, ballad, fretless, and R&B techniques that are sampled through custom-made tube preamps. There is a wide range of variations available with harmonics, glisses, fuzz, trills, slides, and other techniques, as well as effects recorded from both modern and vintage instruments.

The bass synthesizers include high-resolution core samples from analog bass synths like the Minimoog, the Roland Juno 6, the Roland TB-30, the Roland SH-10, the Oberhei SEM, the Moog Taurus, the OSCar, the Virus, the Yamaha CS-80, the Arp Odyssey and 2600, the Studio Electronics SE-1, the Sequential Circuits Pro One, the Moog Voyager, and many others.

Charlie

Ultimate Sound Bank
www.ultimatesoundbank.com

Charlie is a sample-based synthesizer that imitates classic electric organ sounds. There is a very large sound library of organ sounds that were recorded with vintage equipment, including Leslie speakers recorded with low- and high-rotor speed. It covers the classic Hammond B3 sounds and several other popular organs.

Charlie's sample-playback engine has a clearly designed interface and a complete set of effects to shape the sounds (see Figure 7-25). It has unlimited polyphony and there is real-time MIDI control of all parameters.

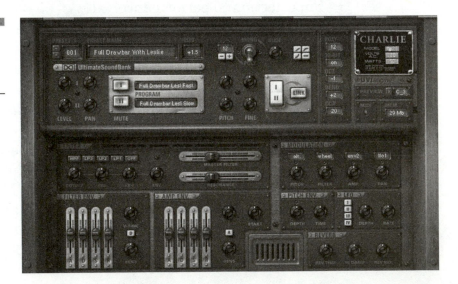

Figure 7-25
Charlie's clearly designed interface

Plugsound

Ultimate Sound Bank
www.ultimatesoundbank.com

Plugsound is a virtual instrument for playing back sampled sounds and is married to a collection of sample-based virtual instruments that includes several volumes, each based on a different professional sample library (see Figure 7-26). The user buys the libraries and the plug-in sound module is included with each volume.

Plugsound has two resonant filters, one multimode and one that is switchable between high pass and low pass. It also includes amplitude and filter envelopes, one LFO that can be used for tremolo, another LFO that provides modulation, auto pan, a reverb module, and MIDI control of every knob and slider on the interface using standard MIDI continuous controllers.

Plugsound Volume 1 is the Keyboards Collection. It has classical, jazz, and pop acoustic pianos with variations, such as honky-tonk and out-of-tune pianos. It also has a

Figure 7-26
Ultimate Sound Bank's Plugsound

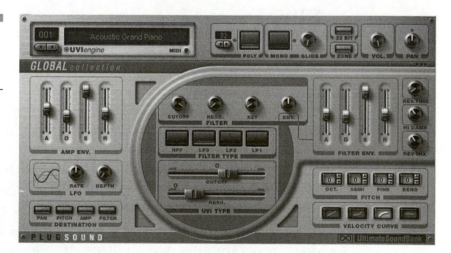

huge array of electric pianos, clavinets, organs, and synthesizer keyboard sounds. Other traditional keyboard instruments include church organs, accordion, celesta, vibraphone, music box, and xylophone. There is also an extensive harpsichord section.

Volume 2 is a collection of fretted instruments that includes virtually all kinds of acoustic and electric guitars and basses.

Volume 3 has over 5,000 drum and percussion sounds, including stylized kits, individual percussion instruments, and digital drums.

Volume 4 has more than 600 presets using 1,500 samples, which are all oriented to hip-hop sounds and compositions. It is full of loops, kits, and multisampled instruments.

Volume 5 is a collection of vintage and modern synth sounds. It has a wide variety of analog, digital, virtual analog, and plug-in synthesizers.

Volume 6 is a wide-ranging collection of samples that conform to general MIDI classifications. They can be used to replace weak sounds with high-quality samples.

Ultra Focus

Ultimate Sound Bank
www.ultimatesoundbank.com

Ultrafocus is a new product that uses a proprietary sound engine and comes with a large collection of sounds created from classic synths. The included presets can be loaded in one of two available layers. Each layer has dual multimode filters, two envelopes, four LFOs, mono-legato mode, and enhanced modulation routing. It uses analog, analog modeling, vector, wavetable, and digital sythesis techniques for deep programming possibilities. It has unlimited polyphony and real-time MIDI control of every parameter (see Figure 7-27).

Xtreme FX

Ultimate Sound Bank
www.ultimatesoundbank.com

Xtreme FX is another new product coming from Ultimate Sound Bank. It shares many features and interface components with Ultra Focus, but is oriented toward

Figure 7-27
The UltraFocus window

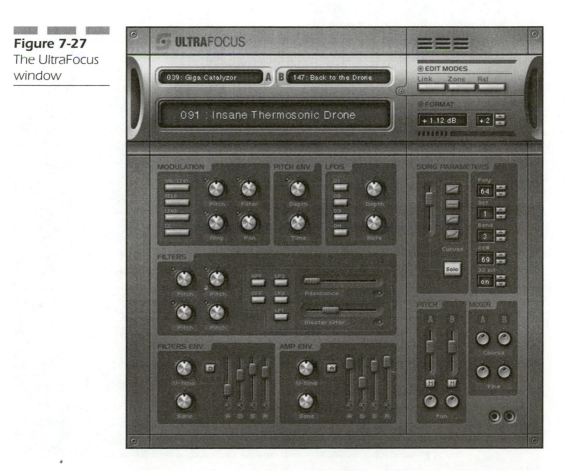

sound design (see Figure 7-28). Each layer has dual multimode filters, two envelopes, four LFOs, mono-legato mode, and enhanced modulation routing. It uses analog, analog modeling, vector, wave table, and digital synthesis techniques for deep-programming possibilities. It has unlimited polyphony and real-time MIDI control of every parameter.

Plug-ins can go a long way toward helping you build a Pro Tools system that will work for your needs. They are not, however, the only outside resources that can add to the program. There are many applications as well that may be useful in your audio-editing projects.

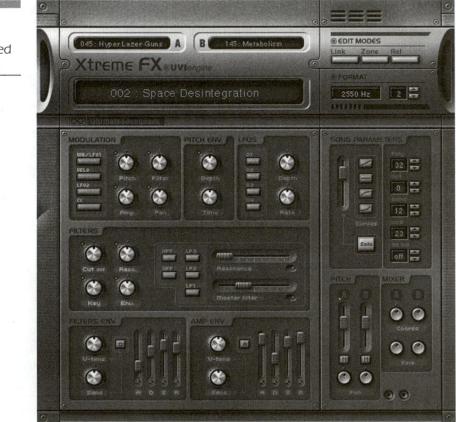

Figure 7-28
The sound-design focused Xtreme FX

CHAPTER 8

Other
Applications

This chapter contains descriptions and explanations of additional software used with Pro Tools. Some of these programs work directly with Pro Tools hardware. Others work in software with Pro Tools files.

Sequencers

There are two major third-party sequencing programs that run on Pro Tools *time division multiplexing* (TDM) hardware: Digital Performer from *Mark of the Unicorn* (MOTU), and Logic Audio from Emagic. They are both very powerful *digital audio workstations* (DAWs) that have many capabilities in common with Pro Tools. Many professionals use more than one of these programs, depending on the type of project they want to do.

Generally speaking, many users consider Pro Tools to be the best program for handling pure digital audio. However, Digital Performer is favored by many musicians because of a number of composer-friendly features. Those that are heavily into *musical instrument digital interface* (MIDI) may prefer the advanced MIDI capabilities of Logic.

Any of these programs can be used for virtually any kind of audio project. Their feature sets are heavily overlapping. However, the internal logic of each program has many differences and each one represents a significant learning challenge.

Digital Performer

Mark of the Unicorn
www.motu.com

Digital Performer is an integrated digital audio and MIDI sequencing production system. It provides a comprehen-

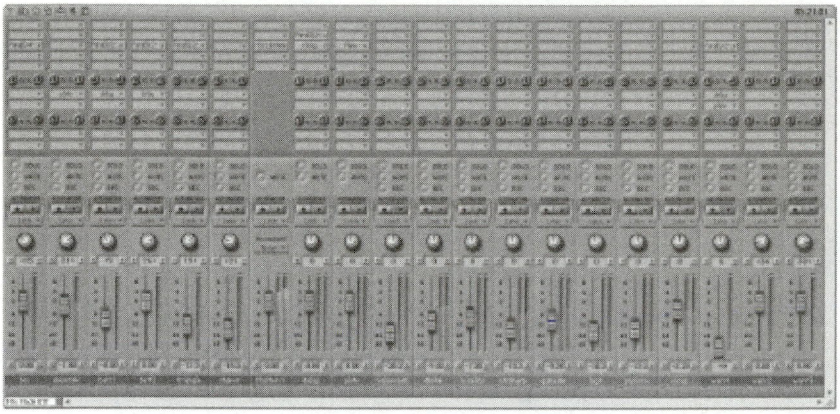

Figure 8-1
MOTU's Digital
Performer

sive environment for editing, arranging, mixing, processing, and mastering multitrack audio projects for a wide
variety of applications (see Figure 8-1).

Audio Digital Performer supports a wide range of audio
hardware, including TDM (Pro Tools), up to high-resolution 24-bit, 192 kHz audio.

Mixing MIDI and audio tracks appear in a single, uni-
fied mixer. Up to 20 effect inserts and 32 stereo buses per
audio channel are available. There are five advanced
automation modes, beat-synchronized effects, and sample-
accurate editing of automation data. The user interface
includes many automation features, such as event flags
for discrete events and spline tools for manipulating control points. Automation parameters are displayed in
meaningful, real-world values such as decibels and
milliseconds, as opposed to arbitrary values (0–127).

Effects Digital Performer includes dozens of real-time
DSP effects, including two-, four-, and eight-band EQ,
tube-simulation, and distortion effects, three reverbs, two

noise gates, two compressors, a synthesizer-style multi-mode filter, and echo and delay effects, including a surround delay, a chorus, a phaser, a flanger, a modulator, and others.

Sample Editing   Digital Performer includes a built-in waveform editor with a pencil tool for removing clicks and a loop tool for defining loops for your sampler. The sample editor allows the drag-and-drop transfer of audio from your project to samplers, which appear as devices inside Digital Performer.

Editing and MIDI   Audio editing is accurate to the sample level. MIDI editing has a resolution of 1/10,000 *pulses per quarter* (PPQ). An unlimited number of MIDI tracks can be recorded simultaneously. Digital Performer has a sequence editor, a graphic editor, an event list, a drum editor, and a notation editor. With the Notation window, you can print out the whole score or individual parts.

The interface uses continuous scrolling, which moves the music under the wiper (playhead), which stays fixed to the center of the window.

Surround Sound   There are four panner plug-ins, including a localizing room simulator. Each audio track can be assigned to any surround sound format—from LCRS up to 10.2. Panning movements can be automated. Digital Performer also has an assortment of variable channel effects processors, including the MasterWorks Limiter. Many composers use Digital Performer as their main *digital audio workstation* (DAW).

Logic Audio

Emagic
www.emagic.de

Logic Audio is a product of Emagic, which was recently acquired by Apple Computer. Some of Logic's technology is beginning to appear in other Apple products, notably Final Cut Pro 4.

Logic is the most modern of the DAWs. It was built after all the others and is, in many ways, a hybrid of many applications. It is a fairly complicated interface that combines composition, notation, MIDI, and audio production facilities (see Figure 8-2).

The audio functionality is very good and Logic is well known for its MIDI capabilities. Its scoring capability

Figure 8-2
The Logic Audio interface is a combination of facilities.

comes from the famous and very powerful Notator program. It has numerous powerful audio tools and excellent compatibility with Pro Tools hardware.

Logic has extensive audio hardware support, including Pro Tools TDM and TDM-HD. It is capable of creating 192 tracks of audio playback at up to 24-bit, 192 kHz rates, in a variety of audio formats. The audio functionality also incorporates a digital mixer with surround sound up to 7.1, over 50 high-quality audio effect plug-ins, and fully automated sample-accurate mixing facilities. It incorporates the high-end POW-r dithering process for mastering. Logic supports an unlimited number of MIDI tracks with timing accuracy to 1/3,840 of a note and it has a wide range of synchronization options.

Logic is the favorite DAW for many heavy MIDI users. It incorporates a technology called *active MIDI transmission* (AMT). It can support really large MIDI configurations with the ability to communicate with hundreds of MIDI ports simultaneously. The program sends and receives MIDI clock, MTC, MMC, and word clock signals, making it ideal for film, TV, and post-production facilities.

The Arrange window is Logic's primary workspace. Audio and MIDI objects are easily resized, soloed, muted, looped, and mixed in real time without stopping the sequencer. MIDI edits, such as quantization, transposition, and note length changes, are also calculated in real time. Markers facilitate jumping to any location, complete with production notes. The program operates on a rack-based automation system and automated parameters are displayed with full names and values.

The *Environment* portion of Logic is an extensive onscreen MIDI environment. It has a flowchart of the connected MIDI equipment and provides diverse options for combining software-based effects, virtual-instrument processing, and other MIDI programs. Custom MIDI controls can be created and used to control both external and internal MIDI devices. Logic's MIDI control is more extensive than in other DAWs.

The user interface in Logic is customizable. You can save up to 90 screensets per song (session). Each screenset recalls the exact screen configuration of every window on the screen.

Each Logic audio track has EQ, eight sends and eight inserts to 64 buses, with inserts and sends also available for audio instruments and live inputs. Insert effects are to be used in inputs, tracks, instruments, buses, and outputs.

Audio units are the program's preferred plug-in format, and there are many available in all the major categories. Up to 64 software-based synthesizers and samplers can be used in Logic's Audio Instrument channels. A few audio instruments are included as standard. The audio instruments are integrated into Logic's mixer, making basically everything available for mixing and processing. ReWire is also integrated.

Session Backup Software

Anyone who has had their computer lock up in the middle of a project knows the value of session backup software. The software available for Pro Tools is generally more sophisticated than a simple auto recovery and the applications perform a variety of functions for the Pro Tools system.

Mezzo and Mezzo Mirror

Mezzo Technologies/Grey Matter Response
www.mezzotechnologies.com

Mezzo is automatic-backup software for Pro Tools sessions. It will back up a session file and also search for all related audio and fade files to transfer them onto the archiving media. Every time the session is saved, Mezzo scans for any changes and adds any new items to a database. After that, Mezzo can restore a single file or a whole session.

The backup and storage activity takes place in the background and is ongoing during a session. A simple interface makes finding and accessing the stored data easy (see Figure 8-3). Mezzo can also store files to multiple devices at the same time. It works with all popular storage devices, including tape drives and disk devices.

Mezzo Mirror allows users to create an exact copy of a project on a separate hard drive or CD/DVD-R to create a plug-and-play backup. All the files are saved in their native format so they are ready to go immediately when they are reloaded. Mezzo Mirror is a bit less expensive than Mezzo and has all the same capabilities, except tape backup.

Live Audio Tools

Live audio tools allow you to take Pro Tools out of the studio and into the world of performance.

Figure 8-3
The Mezzo
scanning
window

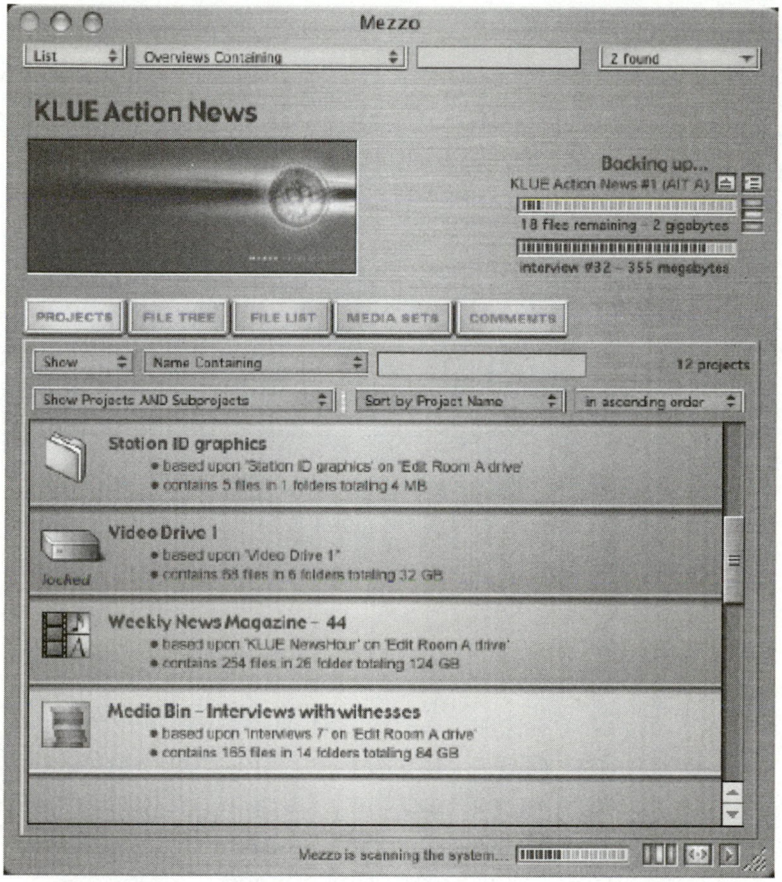

IMEASY

A & G Soluzioni Digitali
imeasy.aegweb.it

The *Integrated Modular and Expandable Audio Spatial-*
ization System (IMEASY) is a professional audio spatiali-
sation system. With IMEASY, you can use Pro Tools as a

Figure 8-4

IMEASY allows you to place speakers in a virtual space to see how they will perform.

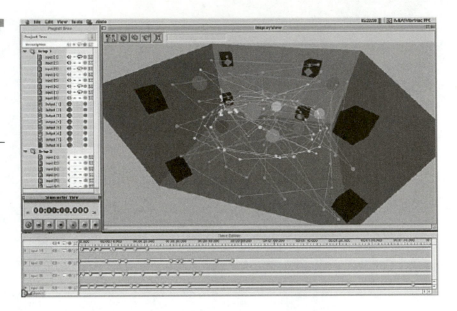

location audio system with up to 24 channels. Speakers can be freely placed with useful control over the sonic space (see Figure 8-4). The program was originally created for live events, but can be used for any kind of spatialized sound environment, such as theaters, theme parks, museums, or other public spaces.

TH-S (Theatre System)

APB Tools
www.apbtools.com

TH-S, a multiplayer, multichannel show-playback system, uses Pro Tools hardware to create four stereo and two eight-track players for live applications with a show-based cue list and snapshot automation, including individual Master Fader levels with CM Automation Motor Mix

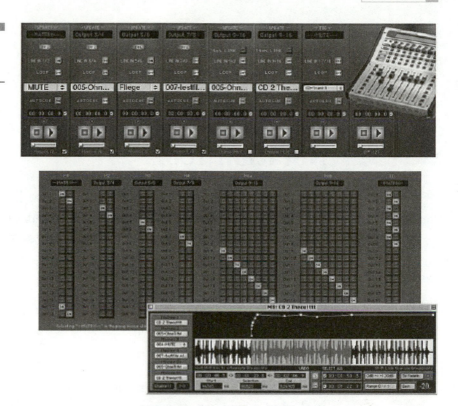

Figure 8-5
The TH-S
interfaces

hardware (see Figure 8-5). It supports up to 16 discrete outputs.

TH-S plays SDII and WAV files directly from the hard drive. The playback takes place from up to 16 outputs, which are routed to the virtual playback machines. Every TH-S player has an individually switchable Autocue mode, in which the player behaves like a tape recorder, in that it stops after the end of the file and is positioned at the next cue. Every sound file starts without delay, as the first sample of every sound file is loaded in RAM.

TH-S also has a loop function. You can cue up and edit new sound files for running players in the background. There is a MIDI program change function for controlling external MIDI sound modules and mixers. In addition, the program can take snapshots of its setups, and a MIDI program-change input can be used to externally recall snapshot setups.

Notation Tools

If you want to be able to recreate your audio projects live, then you may want to invest in a notation application. These tools allow you to put your projects on paper.

Sibelius 2

Sibelius
www.sibelius.com

Sibelius 2 is a full-featured notation software system that includes a variety of components for creating lead sheets and orchestrations (see Figure 8-6). It can read MIDI files from Pro Tools and convert them to scored music.

In Sibelius, new notes can be input directly from the computer keyboard. Percussion instruments can be assigned to keys. There are various tools to streamline orchestration and editing. There are flexible cut, paste, and copy tools; interval and chord tools; and step-time input. Lyrics can be imported from a word processor. You can also scan text or music into the program.

Sibelius is easy to use for simple tasks but has enough features that very complex scores can be handled. The printed ouput looks great.

Figure 8-6
Sibelius 2 creates printable lead sheets.

Postproduction Tools

Postproduction tools will help you in your projects if you are working on audio for a movie or TV production.

MetaFlow

Gallery
www.gallery.co.uk

MetaFlow is an asset-management system for film and TV sound that is designed to handle large-scale audio projects. It has the tools to acquire media assets from a variety of sources, including a tight integration with Pro Tools.

MetaFlow manages the data in a flexible database from which information and files can be retrieved in a variety of manners (see Figure 8-7).

The program can ingest BWF broadcast WAVE files, Avid OMF media, OMF compositions, Avid bins, ALE and FLEx files. It presents Avid AIF and BWF media to Pro Tools users in a bin-style window showing clip names, scene and take, and other Avid bin and OMFI metadata.

MetaFlow also has a multichannel clip editor with multiple waveforms and an onscreen mixer. It can perform automatic spotting of complete multichannel masterclips as a group into a Pro Tools timeline. MetaFlow provides numerous workflow improvements to large-scale audio post projects.

Figure 8-7
The MetaFlow database

Session Browser

Gallery
www.gallery.co.uk

Session Browser allows the user to browse Pro Tools sessions without opening Pro Tools. You can quickly view the session's contents, including the audio regions and dependent sound files, across multiple storage sources. The program will also display plug-in usage for TDM sessions (see Figure 8-8).

The session sample rate converter can turn a 48k session into a 44.1k session and back again, processing all the necessary sound files and the session itself. It can also detect session and file sample rate mismatches and shared media usage.

Figure 8-8
The Gallery Session Browser displays Pro Tools plug-in usage.

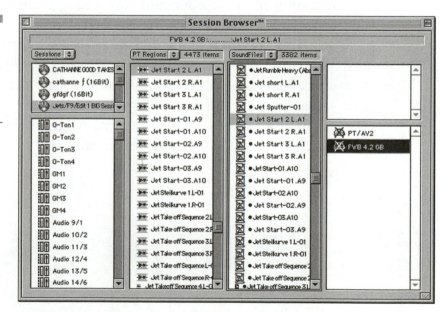

The VariSpeed session function can convert sessions from 24 to 25 FPS, pull sessions up and down, correct sample-rate mistakes, make file names unique, and offer user-definable ratios to correct random sync problems where the session runs slightly too slow or too fast. Session Browser has a number of other features that fix or prevent other session problems common to large postproduction environments.

Virtual VTR

Gallery
www.gallery.co.uk

Virtual VTR turns a Macintosh computer into a stand-alone nonlinear VTR that can be controlled via MIDI or Sony 9-pin protocols, and can also synchronize playback, chasing timecode (see Figure 8-9).

Running on a dedicated *central processing unit* (CPU), the idea of Virtual VTR is to unload the *QuickTime* (QT) video responsibility from systems operating with a QT video card in the same computer as the audio workstation. Unloading frees up system resources on the audio workstation and allows a second computer to be dedicated to high-quality picture playback.

Figure 8-9
The Virtual VTR

Virtual VTR can be used on postproduction mixing scoring, foley, and ADR stages. It provides rapid spotting for ADR and variable speed picture search; it is also useful for TV stations, since it has excellent time-code implementation. The application can record time-stamped live feeds from multiple cameras ready for editing without accessing videotape. It can also perform gang recording of multiple cameras on several synchronized Virtual VTR systems and play them all back in sync. For TV stations, Quick-Time movies can be played from the desktop with a 9-pin and SDI interface for on-air playback or ingest.

Basically, Virtual VTR on a dedicated Mac becomes a high-end SDI nonlinear video-disk recorder that will perform basically any function you would expect.

TITAN

Synchro Arts Limited
www.synchroarts.com

TITAN is a standalone application that batch-processes tracks in a selected Pro Tools session file and either replaces existing tracks or creates new tracks with the desired results.

This application can facilitate the transfer of files between Avid AudioVision and Pro Tools sessions by processing the Avid files into Pro Tools format or vice versa (see Figure 8-10). It maintains sync during conversions.

TITAN also supports multiplexed WAVE, broadcast WAVE, and SDII files. It has full support for multichannel conforming and an unlimited number of audio channels for each EDL track. Flexible filename recognition allows support for a wide range of product-naming schemas. The

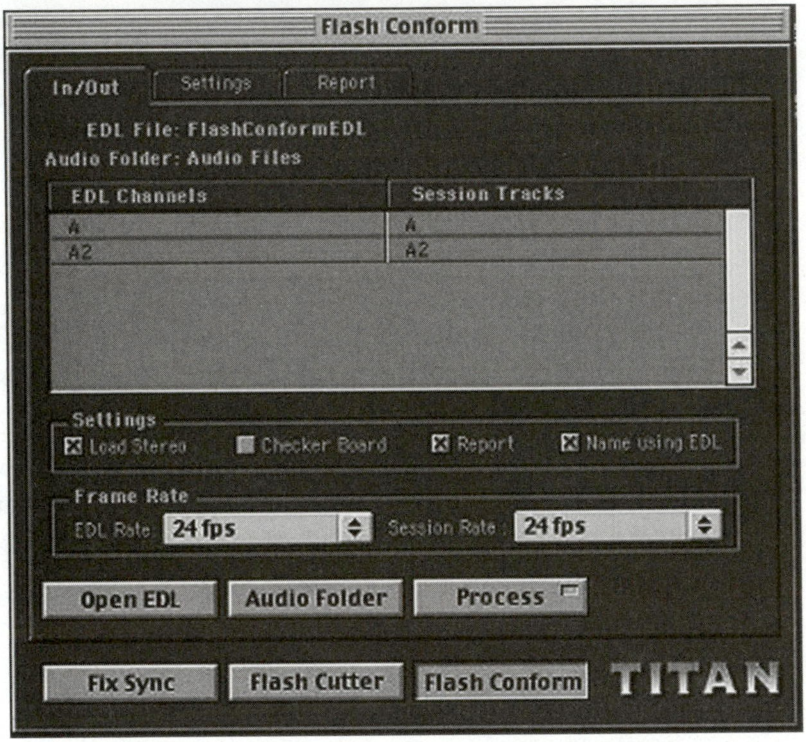

Figure 8-10
The TITAN
window

Figure 8-10
The TITAN
window

application supports reel and channel metadata in broadcast WAVE files. It has audio source file diagnostics and correction, as well as EDL track offset, track disable, and duplicate event removal.

Sound Library Management Tools

The more you work and the more plug-ins and external applications you have, the more tracks you will have to organize. Sound library management tools can help you keep your sounds in order.

NetMix

Creative Network Design
www.creativenetworkdesign.com/

NetMix Pro 2.4 provides real-time multitrack mixing and editing of digital audio streamed from local, network storage, and Internet sources (see Figure 8-11). It can be used as a preview device before importing the audio directly into Pro Tools. NetMix Pro can play, edit, and mix multiple audio files with different file formats from different storage systems simultaneously. The application can also create loops and perform real-time pitch shifts.

Users can prestage a complex sound or music edit in sync to picture using volume, pan, mute, solo-in-place, multitrack scrubbing, waveform editing, and timecode display.

Figure 8-11
The NetMix
Storage System

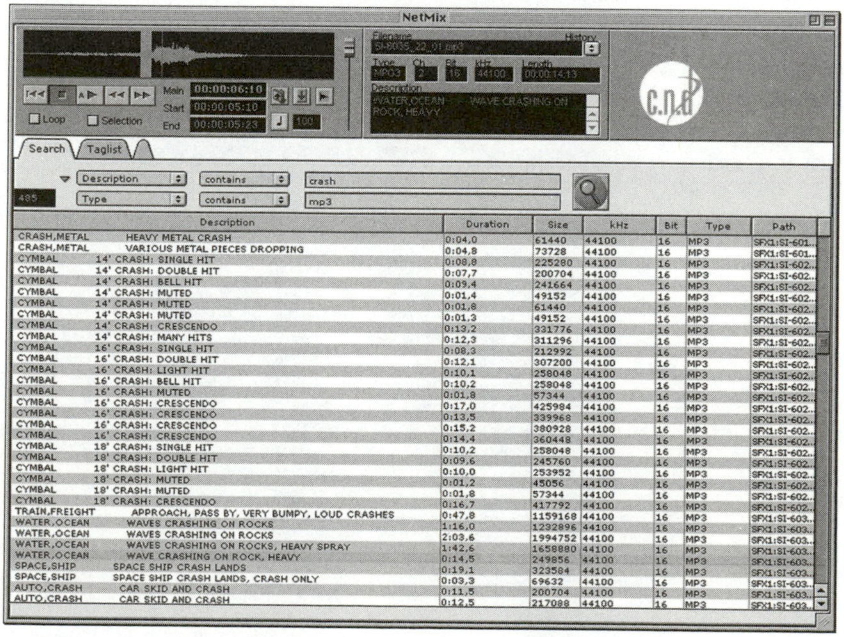

NetMix Pro automatically converts and spots the audio to specific timecodes on the Pro Tools tracks, maintaining all previously made edits.

LibraryLoader

Gallery
www.gallery.co.uk

LibraryLoader is a sound-effects-ripping application for OS X. It can read tracks and indexes directly from CDs and apply names and manufacturers' descriptions for most commercial sound-effects libraries and production music libraries (see Figure 8-12).

LibraryLoader is very fast and can rip an entire CD in as little as three minutes. It is optimized for batch operation. The sound files created by LibraryLoader are recog-

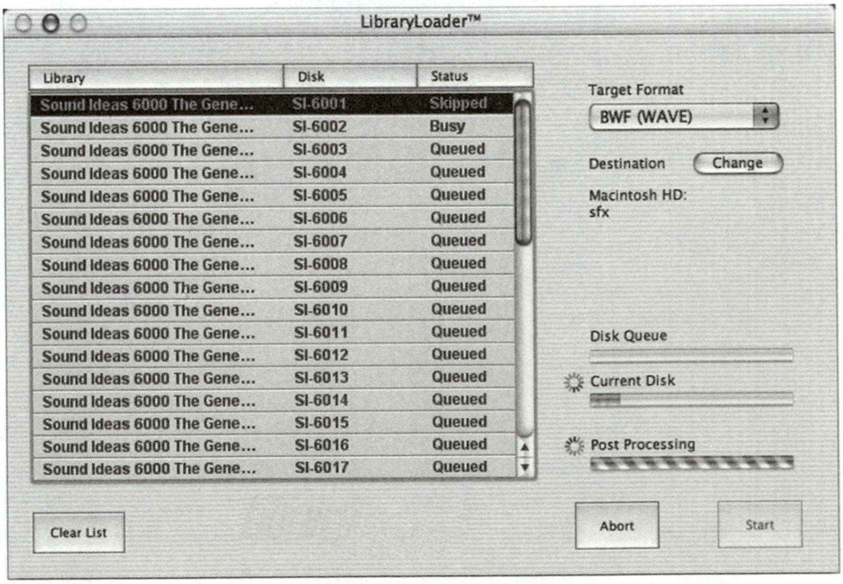

Figure 8-12
Gallery's
LibraryLoader

nized and understood by all major sound-effects-database systems, including Gallery's MTOOLS and Digidesign's Pro Tools 6.0 DigiBase.

Mtools

Gallery
www.gallery.co.uk

Mtools is designed to manage sound-effects libraries of any size with easy accessibility (see Figure 8-13). It can make finding sounds in a large database easy and will automatically transfer them into a Pro Tools session, in the right format and at the right sample rate and bit

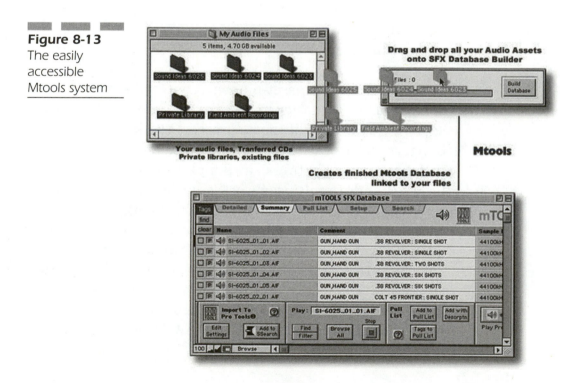

Figure 8-13
The easily accessible Mtools system

depth. The application works in a local machine, on a local area network, or over the Internet.

Mtools converts audio CDs into online, edited file archives, complete with full text descriptions, ready to use in Pro Tools and other audio workstations.

Users can also use MTools to build Filemaker-based MTools databases using a proven commercial database architecture with power, speed, and user flexibility. Filemaker has excellent database management capabilities.

Mtools works with Gallery's own Mtool Server, Appleshare, Windows NT, Unix, or Novell databases. It supports and mixes SDII, AIFF, WAV, or MPEG compressed formats.

Soundminer

Soundminer
www.soundminer.com

Soundminer is an asset-management system for sound designers. It is designed to extract, create databases, search, audition, edit, spot, process, track, and perform batch transfers with speed and accuracy.

Soundminer features a simple one-screen-tells-all browser and a multiuser work environment that can be completely personalized (see Figure 8-14). It performs sample rate and bit depth conversion, as well as waveform overview calculations, on the fly. It has a full set of database tools that automatically check for duplicates and corrupt files. The application supports multiple databases with shared library networking functionality. It works with SDII, AIFF, BWAV, MP3, and AAC (mpeg4) audio file formats.

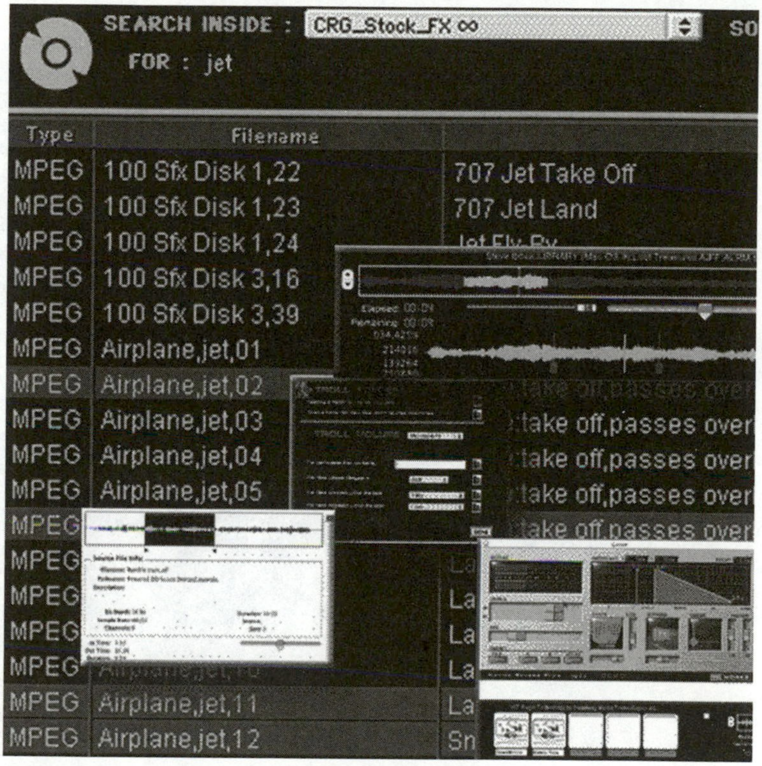

Figure 8-14
Soundminer's
one-screen-tells-
all browser

Soundminer Ripper

Soundminer
www.soundminer.com

Soundminer Ripper is a fast, easy tool that rips audio from
CDs and audio tracks (see Figure 8-15). It extracts track
and index information, prebuilds the waveform overviews,
and embeds all the manufacturer's metadata into the
resulting audio file. Support files for more than 150 popular sound-effects libraries come bundled with Ripper.

Figure 8-15
The Soundminer
Ripper

Users can extract all the track and index information from their sound-effect CDs and then embed the information in their preferred file formats, including Sd2f, AIFF, AAC/MPEG4, and BWAV. When digitizing in BWAV, metadata can be written to make it directly available to Pro Tools 6. Ripper is compatible with DigiBase, and mSoft applications, including network-based ServerSound.

Studio, Project, and Business Management Tools

If you are using Pro Tools for a profession, you may have use for the management tools listed in this section.

Studio Suite v 5.0

AlterMedia
www.studiosuite.com

Studio Suite is a software package for the business and project management of all kinds of media operations (see Figure 8-16). The main menu has 20 buttons that correspond to modules. These 20 modules are divided into three subsets: Office, Studio, and Tech. The application has

Figure 8-16
AlterMedia's
StudioSuite v 5.0

specialized tools for managing recording studios, mastering studios, video postproduction studios, and production companies. It is all written as a set of FileMaker Pro-relational databases.

Studio Suite features built-in relational, cross-platform (Mac/PC), multiuser network capabilities. It is password protectable with five different levels of access and contains a built-in spell checker. It can generate invoices, letters, envelopes, tape labels, and other custom forms.

The application can automatically export data to Quick-Books and other accounting programs. Users can input and manage contacts, calendars, sessions and projects, billing, media, rooms, equipment, repair logs, samples and clips, and custom data. Studio Suite also works with bar code readers for data entry.

Studio Suite requires a copy of FileMaker Pro to run, which is not included.

Pro Tools Hardware and System Configurations

CHAPTER 9

Pro Tools Hardware

Up until about 15 years ago, almost all professional audio was recorded, mixed, edited, and mastered with analog equipment. Tape machines were the only viable recording method, and large analog consoles were the platform for mixing. In the 1950s, the early days of rock and roll, a tape machine with more than a two-track recording capability was rare. Even by the mid-1960s, multitrack analog machines were very limited. For example, the Beatles' *Sergeant Pepper's Lonely Hearts Club Band* album was recorded using four-track tape machines and a lot of track bouncing. Before the development of computer-based systems, editing consisted of bouncing tracks between machines or physically cutting and splicing tape. Mastering was also a tape-to-tape process. There were some digital tape machines on the scene, but they weren't used any differently than analog tape machines. Various digital outboard devices were also beginning to appear, but usually they were patched into analog busses. Routing signals between devices used physical patchbays.

In the 1980s, computer-disk-based recording introduced the reality of nonlinear random-access systems, which provided a whole new way of working with media. Pro Tools was one of the first truly viable computer-based nonlinear systems capable of professional-quality audio work. Today, there are few professional recording studios or audio post-production facilities that do not use Pro Tools systems. It is currently the leading *digital audio workstation* (DAW) for music recording, as well as film and video audio post production. Pro Tools set the professional standard from the beginning of computer-based digital audio recording and production, and continues to do so today.

Following a trend in digital video and audio technology, Digidesign has recently developed less expensive versions of Pro Tools. These systems are designed for home- and project-studio environments and offer many of the advantages and features of the expensive systems typically found in professional facilities. The software and user interfaces are nearly identical throughout the systems. The main differences are in the supporting hardware subsystems and the additional capabilities those subsystems provide.

Types of Pro Tools Systems

There are four types or levels of Pro Tools systems available today. These are the Pro Tools Free system, the Pro Tools LE systems, the Pro Tools *time division multiplexing* (TDM) systems, and the Pro Tools TDM II systems. (There are also a number of older Pro Tools systems, which have been discontinued. See the appendix for a list of legacy Pro Tools systems.) The Digidesign web site provides a lot of detail about the various systems, which we won't duplicate here. The following sections give descriptions of the basic systems and their major features.

Pro Tools Free

The least expensive version or Pro Tools is called "Pro Tools Free." As the name implies, this system is free for download from the Digidesign web site. This version has limited capabilities compared to the other system versions. However, if you have a Macintosh computer or a PC with a sound card, it can provide the opportunity to become familiar with the general concepts of Pro Tools, and if your requirements are fairly simple, it may do the job. If you are interested, check out the Pro Tools Free Appendix at the end of this book.

Pro Tools LE

Pro Tools LE is the next step up from Pro Tools Free. It requires a Digidesign hardware subsystem for input and output. There are currently three types of hardware systems offered by Digidesign that support Pro Tools LE: the Mbox, the Digi 001, and the Digi 002. Each system comes with both the hardware and the Pro Tools LE software. Pro Tools LE systems have all of the features of Pro tools TDM systems, but they are limited to 32 tracks and can't utilize the TDM or TDM II plug-ins. The TDM systems support greater track counts and the real-time TDM/TDM II plug-ins by adding PCI cards to the computer that have dedicated signal-processing microprocessors. The LE systems rely on the host computer for all processing. Pro Tools LE systems range in price from around $500 to around $2,500.

Mbox   The Mbox, a recently introduced system, is currently the least expensive hardware-based system offered. It uses Pro Tools LE software and includes an audio interface box that connects to a Mac or PC *universal serial bus* (USB) port. The interface provides two channels of analog input for microphones or instruments through a pair of Neutrik jacks that accept either *ground left-right* (XLR) or ¼-inch phone plugs. These connectors supply 48 volts of phantom power for microphones and can also accept *high-impedance* (Hi-Z) input signals such as those from an electric guitar. There are also two channels of digital input via a *Sony/Phillips digital interface* (S/PDIF) connector (RCA type). The software will support only two simultaneous inputs at a time, so in most cases you will be using either the analog or digital inputs, but not both at the same time. There are also two ¼-inch analog phone jacks for inserts. Inserts are used to patch an external device, such as a compressor, between the analog microphone or instrument inputs and the Pro Tools recording input. (This patch requires a special Y cable.) The Mbox also has a ¼-inch phone-jack headphone output and a pair of balanced ¼-inch phone jacks for analog line output.

The Mbox is a a high-quality piece of gear. The mic preamps and 24-bit, analog-to-digital converters are excellent quality. The system is also tiny and simple. These characteristics make it good for use with a laptop computer in a mobile system, or with laptop or desktop systems in home studios. The two-channel simultaneous-recording limitation is not a problem for most home studio needs. You can build up a 24-track recording one or two tracks at a time. You can edit and mix the 24 tracks and play them back in a two-channel stereo mix.

There is no *musical instrument digital interface* (MIDI) interface in the Mbox, which means that a MIDI will be required to connect to external MIDI devices with this system.

Digidesign has recently added a new video-interface product called Mojo, which, when coupled with Mbox, provides an excellent low-cost audio postproduction solution. (See "Video Subsytems" for details.)

Digi 001   The Digi 001 is a 19-inch, rack-mount interface unit that has eight channels of analog I/O, an ADAT optical I/O (eight digital channels), and two channels of S/PDIF. It has a PCI card to go in a slot in the computer connected to the interface by cable. There are two inputs on the front panel that use Neutrik jacks and accept either XLR or ¼-inch phone plugs. Each of these inputs has a mic preamp, a gain control, a -26 dB mic-pad switch, and a 48-volt phantom power switch. There is also an output gain control and a ¼-inch headphone-out jack with an independent gain control. The rest of the I/Os are on the back of the Digi 001 rack. The rear panel analog Inputs 3 through 8 use ¼-inch TRS jacks and accept balanced or unbalanced inputs. There are ¼-inch TRS balanced/unbalanced outputs for each of the eight analog channels, and another pair for monitor outs. There is a ¼-inch footswitch jack as well for using the system's Quickpunch feature.

Unlike the Mbox, the Digi 001 does provide one MIDI-in and one MIDI-out. There is also a 9-pin serial port for connecting to the host computer's serial port for MIDI control. On modern Macs, this connection will require an adapter cable and an adapter to connect to the Mac's USB port.

In addition to the MIDI support, the Digi 001 can record more channels simultaneously by utilizing the additional inputs. The Digi 001 is now considered a Digidesign legacy product and there are many of them in use.

Digi 002   The Digi 002 is the latest addition to the Pro Tools LE family. It is the most powerful of the Pro Tools LE systems. The hardware in this system consists of a single unit that connects to the host computer via a FireWire cable. It is a combination of a control surface, audio I/O, and MIDI I/O. The control surface has eight motorized touch-sensitive faders, eight motion-sensitive rotary encoder knobs, 10 LED alphanumeric displays, some transport controls, and several other button controls and indicators. Most of these are soft controls that change function depending on the task being addressed. For instance, the rotary encoder knobs can be used to control pan, sends, or plug-in controls, or the string of lights surrounding the knobs can act as level meters. Digi 002 is a well-thought-out and efficient design. The system can also function as a standalone eight-channel digital mixer with built-in EQ, dynamics, and effects. It has eight channels of analog I/O and four of these channels have microphone preamps. There are two channels of dedicated monitor output and a headphone output with a separate volume control. It also has ADAT optical in and out (eight digital channels) and S/PDIF in and out (two channels). There is also an unbalanced input with RCA connectors for CD or tape players. It has one MIDI-in port and two MIDI-out ports. The A/D and D/A converters can handle sampling rates of up to 96 kHz. With the Pro Tools LE software and all the inputs and outputs connected, Digi 002 can handle 18 channels of simultaneous I/O.

The Digi 002 is a very flexible unit and the FireWire connectivity to the host computer ensures adequate data rates. Also, a second FireWire port on the unit allows direct connection of a FireWire storage drive.

Pro Tools TDM and TDM II Systems

Pro Tools TDM and TDM II systems are the high end of the Pro Tools family. The Digidesign innovation TDM first appeared in the early 1990s. It is a high-performance effects-processing and signal-routing architecture that uses dedicated chips on PCI cards, instead of the host system's main processor, for processing. This extremely powerful processing allows Digidesign and third-party plug-in manufacturers to design more complex and demanding software that can perform complex tasks in real time. This capability opened the door for some very powerful—and often better sounding—effects, and gives the whole system a real-time feel.

Another advantage of this system is that it means that more tracks can be added by adding more PCI cards, and each of these cards can support more simultaneous channels of *input/output* (I/O). TDM systems can be expanded to allow 64 tracks of simultaneous I/O.

TDM II, which was recently introduced, is a redesigned and far more powerful version of TDM. It is the technology used in the HD systems. TDM II doubles the simultaneous I/O to a possible 128 tracks, and at least doubles all the other capabilities of TDM systems. HD systems can handle sampling rates up to 192 kHz.

Unlike the Pro Tools LE systems, which are complete, ready-to-go software/hardware systems, TDM and HD (TDM II) systems are highly modular custom configurations built from several PCI cards, external I/O units, various special function external modules, and a variety of control surfaces. In addition to Digidesign, many third-party companies make compatible hardware accessories. Pro Tools TDM/TDM II systems can range in price from about $8,000 to upwards of $70,000 or more for a large-scale system with lot of accessories.

When you design a TDM or HD system, you begin with a core system consisting of the Pro Tools TDM software and one or more PCI cards.

TDM Core Systems

The only current core TDM system available is the Pro Tools 24 MIX. This system has been the leading professional system for a number of years, and there is a large installed base of these systems in recording and postproduction facilities around the world. Currently, many of these facilities are replacing their 24 MIX systems with the Pro Tools HD system, and this trend will continue. However, the 24 MIX systems are robust and I expect many will be in operation for years to come. MIX systems are capable of working with sampling rates of up to 48 kHz.

There are two types of PCI cards in MIX systems: the MIX Core card and the MIX Farm card. All MIX systems have a MIX Core card. It is the main Pro Tools processor. The Mix Core card requires at least one I/O selected from several that are available. It can support up to 16 channels of simultaneous I/O. One or two MIX Farm cards can be added, each providing the capability of 16 more channels of simultaneous I/O and additional *digital signal processing* (DSP) chips.

The Pro Tools 24 MIX system has a single MIX Core card and can provide up to 16 channels of simultaneous I/O. The Pro Tools 24 MIX Plus system has a MIX Core card and one MIX Farm card and can provide up to 32 channels of simultaneous I/O. The Pro Tools 24 MIX^3 system has a MIX Core card and two MIX Farm cards and can provide up to 48 channels of simultaneous I/O. They can all play and record up to 64 tracks through their available I/O. Additional MIX Farm cards and I/Os can be added to increase the number of I/Os and available DSP.

Input Output Interfaces for MIX Systems The MIX Core and Mix Farm cards can be connected to four different types of I/O interfaces. At least one interface is required for the system to work. The interfaces are all rack-mount devices. In addition to their varying I/O configurations, the cards also all have word clock-in and clock-out for synchronization. (See the section on synchronization for more information.)

888-24 I/O   The 888-24 I/O has eight balanced analog channels of I/O, each with XLR connectors and high-quality 24-bit *analog/digital* (A/D) and *digital/analog* (D/A) converters. It also has eight channels of 24-bit digital I/O with XLR connectors for direct 24-bit digital transfer to and from digital sources. The interface also has two channels of S/PDIF I/O.

882-20 I/O   The 882-20 I/O has eight balanced analog channels of I/O, each with ¼-inch TRS jacks and 20-bit A/D and D/A converters. It also has two channels of S/PDIF that operate at 24 bits.

1622 I/O   The 1622 I/O has 16 balanced analog input channels of input, each with ¼-inch TRS jacks and 20-bit A/D converters. The interface has two 24-bit channels of analog output with ¼-inch TRS jacks. It also has two channels of S/PDIF that operate at 24 bits.

24-Bit ADAT Bridge I/O   The 24-bit ADAT Bridge I/O has two sets of ADAT optical I/Os, each supporting eight 24-bit digital channels. This interface can communicate with any device that supports the ADAT protocol, including a number of digital mixers. It also has two channels of AES/EBU digital I/O using XLR connectors, two channels of S/PDIF, and a pair of balanced ¼-inch TRS analog outputs.

HD Core Systems

The basic architecture of HD systems is similar to that of MIX systems. However, it incorporates a number of advances and provides a much higher performance, both in the number of tracks supported and in the amount of DSP-based processing power. The general consensus is that the audio quality and overall performance of HD systems is much improved over that of MIX systems.

As with MIX systems, there are two kinds of PCI cards for HD systems—HD Core cards and HD Process cards. All HD systems begin with one HD Core card and require at least one interface.

The HD1 system uses one HD Core card only. It can support up to 32 channels of I/O and can play 96 tracks of audio. The HD2 system has an HD Core card and one HD Process card. It can support up to 64 channels of I/O and can play 128 tracks of audio. The HD3 system has one HD Core card and two HD Process cards. It can support up to 96 channels of I/O and can play 128 tracks of audio. The system is expanded further by adding additional HD Process cards and interfaces as needed.

HD Interfaces

192 I/O The 192 I/O can support up to 16 channels of analog and digital input and output through a variety of possible connections. Using DB-25 connector and breakout cables, there are connections for eight channels of analog I/O, eight channels of AES/EBU, eight channels of TDIF, and 16 channels of ADAT. There are two additional channels of AES/EBU with XLR connectors and a two-channel S/PDIF digital I/O. It works with sampling rates of up to 192 kHz.

The 192 I/O has an additional I/O option bay that accepts cards that deliver either eight additional channels of analog input, eight channels of analog output, or eight channels of AES/EBU, TDIF, and ADAT I/O connections.

192 Digital I/O   Also supporting sampling rates of up to 192 kHz, the 192 Digital I/O has no analog inputs or outputs, just digital connections. It provides up to 16 channels of AES/EBU and TDIF I/O using DB-25 connectors. It also has 16 channels of ADAT I/O, along with two channels of S/PDIF I/O.

96 I/O   The 96 I/O, the least expensive HD interface, supports sampling rates of up to 96 kHz. It has eight channels of analog I/O with ¼-inch TRS jacks, eight channels of ADAT optical I/O, two channels of AES/EBU, and S/PDIF I/O.

All of the HD interfaces have word clock-in and clock-out for synchronization. They also all have a *legacy peripheral* port that facilitates connecting one on the MIX interfaces for additional inputs and outputs. However, these will only support sampling rates up to 48 kHz. They also include 9-pin serial ports for connecting various accessories.

Building and Expanding the System

Most Pro Tools systems have at least some additional hardware involved. The extent to which you will need to expand upon the basic system depends on what system you have and what you want to accomplish with it.

At the most basic level, Pro tools LE systems can be fairly simple. However, it doesn't take long before certain additions to your system become desirable.

Computers

For some reason, the evolution of personal computers has led to both Mac and PC users having very strong feelings about their platform choice. Most users are strongly committed to one or the other, and are often willing to defend their choice quite emotionally. These commitments sometimes tend to be a bit irrational. But in actuality, most users are more comfortable with whichever platform they are used to and know the best.

So, although any discussion of the choice of a Macintosh or a PC for Pro Tools systems may run the risk of offending one's techno-religion, there are a few things to think about when deciding on a platform for Pro Tools.

Factors to Consider   Pro Tools was originally designed for Macintosh and has since been ported to PC. Most of the high-end systems are still based on Macs. The latest version of Pro Tools, Pro Tools 6, was first made available for Macintoshes running OS X. (Version 6.1 will support both Mac OS X and Windows XP-based systems.) Although Digidesign is committed to supporting both platforms, they, like all companies, have limited resources. Since the majority of high-end systems are Mac based, it would be logical that they will continue to develop first for Macintosh and then release PC versions to keep up with the changes made on the Mac versions. This tendency would seem to suggest that, if you are putting together a high-end system, the Mac would be the best choice. Other considerations include the fact that there are more plug-ins available for the Mac than the PC, and there are more trained operators who learn on a Mac. The latter trend could be a consideration if you plan to run a facility and will need to hire operators.

The downside is that Macs tend to be more expensive than PCs. However, this tendency may not be a serious consideration, as with a high-end system, the cost of the platform is a small fraction of the overall expense.

If the other software you are using or want to use is PC based, you may consider a PC-based system. If you are a heavy ACID user, for instance, you may want to stick with the PC.

The downside of PCs is that they have many variations in their configurations and their operating systems. Macintosh systems and operating systems all come from the same place and the degree of variation between models is far less than the variation among PCs. System configuration and maintenance tend to be easier with a Mac.

Although most of the MIX and HD systems are Mac based, about as many Pro Tools LE systems are sold to PC users as Mac users. If you are going to be using a Pro Tools LE-based system, you will not be bucking any trends by installing it on a PC. In the case of Pro Tools LE, the cost of the computer system is a significant part of your total purchase. If you already have a computer that is adequate, then there is no immediate need to purchase another one. Once again, compatibility with other software that you want to use is also an important factor.

Macintosh OS 9-based computers tend to crash quite often. However, this problem has been solved with OS X, which is a very stable operating system. The worst of the problems associated with early versions of OS X have been solved, and it is a good choice at this time. Pro Tools 6.1 is the first version of Pro Tools to be written for Windows XP. Although it may be stable as soon as it is released, it is also possible that there will be a few rough spots at first.

Digidesign, like all good developers, will be right on top of this and will no doubt deal with any problems that come up. However, software developers are often at the mercy of the operating-system companies, who are famous for making late-breaking changes and frequent updates that can wreak havoc on complex applications such as Pro Tools.

Whichever system you decide on, you want the fastest processors and as much RAM as possible. Pro Tools is processing intensive and the more robust the host machine is, the better overall performance you will have. AudioSuite and RTAS plug-ins rely on the host processors for their operations and, if you use a lot of them, the capability of the system will be critical.

Hard Drive Storage Just a few years ago, adequate hard drive storage for digital media was expensive and cumbersome, but the current trends in hard drive performance and pricing present many attractive choices for Pro Tools users. Although recording media to your system drive is possible on some systems, it is generally not a good idea. The constant writing and erasing of large media files results in eventual disk fragmentation. Defragmenting or reformatting disks is a part of regular maintenance on media drives and will be much more complicated if your system disk is involved.

One or more separate drives should be used. As you add more tracks and work at higher sampling rates, the size and performance requirements of hard drives increase.

As a general rule, audio drives should operate at a minimum of 7,200 *revolutions per minute* (RPMs) and be able to sustain data read and write rates of a minimum of 3MB per second. They should also have a large capacity. The higher your sampling rate, bit depth, and track count, the more disk space you will need. The storage requirements for one minute of a single track of audio ranges from 5MB at 44.1 kHz/16 bits, up to 34.4MB at 192 kHz/24 bits.

If you are going to be playing back video simultaneously, it is best to dedicate one or more drives exclusively for that purpose. The disk-speed requirements for video far exceed those for audio, and playing back video and audio from the same drive can result in performance problems.

USB drives are not fast enough to be used for audio applications. However, with that exception, most other kinds of drives can be used, depending on the specific application.

Internal ATA/IDE Drives   The internal ATA/IDE drives of most Macs and PCs are usually ATA (IDE)-type drives. On older systems, these tended to be too slow for recording and playing back audio. However, the internal drives in most newer systems do have adequate speed for recording. Typically, with most desk-side tower computer configurations, including both Macs and PCs, there is room inside the tower for one or more additional internal drives. In most cases, there will also be additional pre-wired connections to the system's power supply and an IDE bus ribbon cable with additional connectors ready for connection to an additional drive. If this is the case with your host computer, installing an additional drive is a relatively simple matter. If you are not planning on transporting sessions in progress to other systems, this method is a good low-cost alternative.

SCSI Drives   SCSI drives have been the most used type of drive for Pro Tools systems in the past (before IDE and FireWire drives reached adequate performance levels). They are generally the most expensive type of drive, but offer the highest performance of any currently available drive. For HD systems, SCSI drives are recommended because of their high throughput.

SCSI drives require a SCSI controller. Neither current Macs nor current PCs have any SCSI controllers built in. Single- or dual-bus SCSI controller cards are typically added to a PCI slot and special multipin SCSI cables connect the controller card to the drive.

Configuring SCSI drives can be tricky. Up to seven SCSI drives can be daisy-chained on an SCSI bus (sometimes called a channel). Each drive must be assigned a unique number, called an SCSI ID, which is a physical setting, usually of jumper settings on the drive itself or a switch on an external drive's housing. The total length of an SCSI bus is limited to relatively short runs and varies somewhat by the types of drives and controllers in the system. The cables and connectors are fragile. Special drive-utility software is often required for configuration and maintenance.

Removable Drives Some SCSI systems use drive bays attached to the computer and removable drives. This technique makes the drives portable without having to disconnect and reconnect the cables. Removable drives are useful if you are using a system for multiple clients and want to keep their media separate, or if you are moving projects and media between similarly equipped systems.

FireWire Drives FireWire drives are a modern innovation in drive technology and, for many applications, are the easiest to work with. They are inexpensive and, typically, small and portable. They are plug-and-play devices and require no special configuration. FireWire drives are a good solution for backups and for transport from one system to another, especially since they are compatible with most systems without adding any special drive bays. They are fast enough for most Pro Tools uses, with the possible exception of high-resolution, high-bit-rate HD.

All FireWire drives are actually IDE drives that are connected to the FireWire bus by means of IDE-FireWire bridge-and-controller circuitry, and are usually housed in an enclosure. There are various kinds of these controller circuits. For use with Pro Tools, it is important that the drive is compliant with the Oxford 911 standard. These compliant drives can be hot-swapped by a single cable that plugs into the computer. FireWire drives require that the computer be equipped with a FireWire port. All modern Macs have FireWire ports built in. Most PCs require a FireWire PCI card. There are some systems, notably the Glyph Trip 2 system, that provide a bay attached to the computer with removable FireWire drives.

RAIDs   A RAID is a set of two or more SCSI or IDE drives that share data handling and are connected to a RAID controller. Although not required for most audio applications, RAIDs are a useful alternative for large-scale applications that need massive throughput.

There are several types of RAIDS or RAID levels. The most common for audio and video is RAID Level 0. In a RAID Level 0 configuration, the data is striped across two or more drives. This means that, as the data flows to the disk array, it is broken into small blocks of data and written to separate disks. When it is played back, it is reassembled into the original data configuration. With each disk operating near its peak performance, the overall read/write speed achieved is far greater than with a single comparable disk. As you get into large-scale Pro Tools systems that are writing and reading large numbers of high-resolution tracks or video files, RAID Level 0 performance becomes the best alternative. The downside of RAID 0 is that if one drive in the array fails, all data is lost. There is no way to rebuild the lost data.

RAID 3 and RAID 5 are two other types of RAID configurations that are sometimes used with media applications. They represent two different methods of using a parity configuration that essentially provides an internal backup. If a drive fails, the system can recover the data because it has been backed up on parts of other drives in the array. This method requires more overall disk capacity and can be slower than RAID 0. But in some applications, where data protection is of the highest importance, these types of RAIDs can be a solution.

Networks, Fibre Channel, SANs, Servers, Avid Unity In larger facilities and workplaces that have multiple Pro Tools systems, and possibly other digital audio workstations, video systems, or other media systems, large-scale networks and storage systems are often employed to facilitate the efficient handling of data and media. There are many different types of systems and concepts for facility design that involve various high-speed networks, shared storage systems, and software that manages both the systems and the media.

Newer Macintoshes come equipped with Gigabit Ethernet network interfaces built in. On PCs, Gigabit Ethernet can be added by installing a *network interface card* (NIC) in a PCI slot. A simple Gigabit Ethernet can be established, simply by attaching the computers to a Gigabit Ethernet switch and some minimal software setup. One machine with substantial storage can act as a server, which can store small files, such as sound libraries, which the other machines can access when needed and transfer to their local disks.

This method becomes cumbersome if you need to move large amounts of data often. Moving large media files, in particular, can be very slow. This is where the concept of shared storage comes in. SANs are a technology in which several workstations are connected via a network to a shared storage device. Multiple computer workstations, called hosts, can access the same stored data (media), without having to move it over a network to a dedicated set of local drives. Software manages the flow of the data and keeps track of versions of data that may have been worked on by different hosts.

For audio/video SANs, Fibre channel networks have become an accepted technology. Fibre channel has several advantages over SCSI for these kinds of applications. It can be faster, connect more devices, and cover greater distances than SCSI.

Avid makes a set of sophisticated SAN networking components and systems, including storage drives, switches, and the Avid Unity Media Network File Manager, which is designed specifically to support Avid and Pro Tools facility operations. There are several other companies that make SAN products designed for media as well, including Rorke Data and Medea.

Monitors Quality audio monitors are an essential part of any Pro Tools system. The idea is to use high-quality monitors that let you hear exactly what your system is producing without any distortion or coloration. When you are mixing, you are completely dependent on the sound coming from your monitors. You cannot mix what you cannot hear.

There are basically two types of monitors, passive and active. Passive monitors require an external power amplifier. Active monitors have power amps built into their enclosures. Active monitor technology has become very advanced in the last few years and is a good choice for most applications. Active monitors offer economy and simplicity. The built-in power amps and frequency crossovers are tailored to the specific speakers, creating efficient and sonically correct systems. Mixing environments typically employ more than one set of speakers of different sizes and at different distances from the listener. Being able to check a mix on different systems can reveal problems and help in attaining the best overall mix.

Stereo monitors should be positioned properly. If they are too close together, you will not be able to hear the stereo definition. The sound will be smeared together. If they are too far apart, the focal point of the sound will be behind you, and you will be sitting in a dead space. The basic starting position should be with the monitors about as far apart as the distance from your listening position. You can tweak the arrangement from there, so that you can hear both channels clearly.

Surround-missing environments require a subwoofer and five to seven matched, full-range speakers, depending on the type of surround mix. The most common type of surround sound is 5.1. In this configuration there are six distinct speakers. There are three across the front, two in the rear corners, and a sub-woofer. The sub-woofer is the .1 in the 5.1 and is also called the LFE or low-frequency effects channel. Most consumer 5.1 surround systems have a bass-management circuit that takes all the frequencies below 80 Hz from all the channels and routes them to the sub-woofer. It is critical when doing surround mixing to reproduce this phenomena in the control room. It is the only way to hear what the end-user will actually be hearing.

There are many other variations of surround sound used in specialized theater systems and in custom installations, such as theme parks and other location sound systems. Pro Tools can be configured to emulate these environments by routing outputs to as many channels as necessary.

Speaker placement for surround sound does have standards, but it is actually more forgiving than working with stereo sound. With surround sound, the whole environment becomes permeated with sound and there is less of a sweet spot where the sound is focused.

Video   If you are going to be working with audio for video or film postproduction, it will be necessary to add video capabilities to your system. There are several types of video subsystems that can be used with Pro Tools that will enable synced video playback for audio postproduction.

FireWire Video SubSystems   With Pro Tools LE or TDM systems, video can be imported as a QuickTime movie and placed directly into the Pro Tools timeline. The video will play back in a QuickTime player on the computer screen, but this is generally not adequate for seeing the picture and synchronizing your edits with it. With Pro Tools 6, you can route the video out through one of the computer's FireWire ports. From there, you can route it through a DV-to-composite video converter, such as the Canopus ADVC 100, and then to a video monitor. The monitor should be positioned directly in front of your listening position. When this is set up properly, you will see the current frame of video at the playhead position at all times.

AV Option XL Beginning with version 6.1, Pro Tools MIX or HD systems are also compatible with the AV option XL. The option includes the Avid Meridien PCI video card, which is the same card used in many Avid video-editing systems, including Avid Symphony, Media Composer XL, and Express NT. It also includes a rack-mount breakout box for video I/O and AV option XL software.

With AV option XL, an Avid-based video project can be imported into Pro Tools with the full timeline information intact. You can actually see the video edits and clips, and use the video edit points to navigate the Pro Tools timeline. This provides a high degree of efficiency and accuracy when spotting sound edits. With AV option XL, you can also capture video clips and make some video edits while in the Pro Tools system. This system is typically found in high-end audio postproduction facilities.

The Latest Video Options In April 2003, Avid/ Digidesign announced a new product line called the DNA family of hardware. These products will provide two new video options for Pro Tools users: Mojo and V10.

Mojo The Mojo is a portable video converter/accelerator and I/O device that is smaller than most notebook computers. It is designed to provide real-time I/O of DV, S-video, and composite video. It connects to a notebook or tower Mac or a PC via a single FireWire cable. It can be genlocked (synced) with other devices, including Pro Tools LE and TDM/TDM II systems. It will support all Avid formats, including 24-frame video.

AV Option V10 The newest Avid-based high-end video option from Digidesign will be the V10. It uses the next generation of Avid video hardware and will provide playback of uncompressed 10-bit standard-definition video over SDI. It will have video-capture and conform capabilities and be fully compatible with Avid's new Adrenaline line of media composers.

Controlling External Video Devices   The transports and some functions of external video tape decks and disk recorders can be controlled directly from Pro Tools MIX and HD systems. This activity requires either the Universal Slave Driver (MIX systems) or the Sync I/O (HD systems) and Machine Control (an additional Digidesign software module). These devices connect to professional videotape and disk devices that support RS-422 machine control and SMPTE time code protocols via a 9-pin DIN serial cable. See the following section on synchronization for more information on these devices.

MIDI   MIDI is not audio. It is a control signal that tells connected devices what to do. Technically, MIDI is a serial data protocol that is used to send and receive control information between external interfaces, controllers, sound modules, mixers, control surfaces, internal computer-based software MIDI programs and sound modules, and other MIDI devices.

If you are going to be using MIDI devices with Pro Tools, you will need some way of getting the MIDI information in and out of the system. MIDI hardware devices have 5-pin DIN jacks, and MIDI cables have matching 5-pin DIN plugs.

Neither computers nor Pro Tools MIX and HD system hardware have built-in MIDI ports. The Digi 001 has one MIDI-in jack and one MIDI-out jack. The Digi 002 has one MIDI-in jack and two MIDI-out jacks. All other Pro Tools systems will require a separate MIDI hardware interface. There are many versions of these devices available from a variety of manufacturers. They come in configurations ranging from one MIDI-in and one MIDI-out, up to eight MIDI-ins and eight MIDI-outs. In some cases, multiple MIDI interfaces can be linked to provide even more MIDI ports. There are two basic ways the interfaces connect to the computer, either by older-style serial ports or modern USB ports.

Serial MIDI   On older Apple Macintoshes, there were dedicated serial ports (the printer and modem ports). These ports were used to connect to external MIDI devices that had the MIDI 5-pin DIN connectors. All Macintoshes since the blue and white G3 towers have shipped without these serial ports. The functions once performed by those ports are now performed by the more modern USB ports that are on all Macs.

On PCs there are several ways that MIDI devices can be connected. Some sound cards have MIDI ports built in. Most older PCs have parallel or serial ports that can accommodate a variety of adapters and MIDI interfaces.

At this point, using serial ports for MIDI is an obsolete solution, since Pro Tools 6 and 6.1 require computers running OS X or Windows XP. All of these newer systems will be equipped with USB ports, a much simpler and more reliable solution.

USB MIDI   USB is the modern approach to low data rate serial communications on both Macs and PCs. Current MIDI interfaces use this technology as well. In most cases, the MIDI interface connects directly to one of the computer's USB ports.

The MIDI interface will have multiple MIDI ports. Your MIDI instruments, controllers, sound modules, mixers, and control surfaces are connected to the interface via MIDI cables. In the case of the DIGI 001 and DIGI 002, MIDI devices can be connected directly to their built-in MIDI ports.

There are also a growing number of MIDI controllers that plug directly into a USB port. They have internal MIDI interfaces and do not require any MIDI cabling or connection to an external MIDI interface.

Practically speaking, for most Pro Tools systems, the available number of USB ports on the computer will be inadequate. Most Macs have two USB ports. Once again, this number varies with PCs. One Mac USB is taken by the keyboard. If you have plug-ins that use the iLok authorization key—a USB device—those will use the second one. You may want to add a printer, dongles for other programs, a USB-connected control surface, or other USB devices in addition to your USB MIDI interface. The solution is to get a powered USB hub that plugs into one of the computer's USB ports and provides connections for four to eight USB devices. The reason to get a powered USB hub is because some USB devices draw power from the USB connection. Without sufficient power, some devices may not work. The power provided from the computer's USB port will not be adequate when it is split between multiple devices on a hub. Powered hubs provide a much higher degree of reliability.

MIDI Controllers and Devices   All MIDI hardware devices have some combination of MIDI-in, MIDI-out, and *thru* connectors. Most have all three. Each MIDI port attached to the computer supports 16 channels of MIDI information. If you have only one set of MIDI-in and

MIDI-out ports connected to your computer (as in the case of the Digi 001), you are limited to 16 total MIDI channels. Each additional set of in and out ports allows for another 16 channels. For instance, using an 8 × 8 MIDI interface (eight MIDI-ins, eight MIDI-outs) can handle 128 MIDI channels.

Some MIDI devices use a single channel, such as a wind controller, which is played like a saxophone and designed to be monophonic (one note at a time). Other devices can use multiple channels. These would include polyphonic keyboard MIDI controllers or a MIDI guitar controller set to use a different MIDI channel for each string.

Multiple MIDI devices can be daisy-chained by using MIDI-thru ports. The MIDI-out from the computer connects to the MIDI-in of the first device. The MIDI-thru from the first device connects to the MIDI-in of the second device and so on. In this way, the MIDI information from the computer can drive the various connected devices. For the connected devices to act as controllers with Pro Tools, the MIDI-out port of the device must be connected to a MIDI-in port of an interface connected to the computer (including the MIDI-in on the Digi 001 or Digi 002), or have a direct USB connection.

The MIDI data, because it is serial in nature, flows through all the devices connected to a port. MIDI channels are assigned in software or firmware on both the sending and receiving devices. The devices will respond only to the data intended for them on their assigned channel(s) and ignore the data on other channels, even though it may flow through their MIDI ports on the way to another device.

MIDI information is not sound. For most external MIDI devices that produce sound, it is necessary to route the audio outputs of the device to a Pro Tools audio input. In this way you can record the audio output of the device into a Pro Tools audio track. If the MIDI instrument is a virtual instrument internal to the computer, it can generally be routed internally to Pro tools. There are different ways to perform this routing, based on the particular software instrument, which are addressed later in the book.

Synchronization Hardware   Synchronization is the process by which the recording and playback of different devices can be locked together and forced to run in time with each other. There are external devices designed to facilitate synchronization for Pro Tools MIX and HD systems. Typically, it is necessary to synchronize when you want to run Pro Tools in with external video decks, video-disk recorders, audio-tape recorders, or audio-disk recorders.

For MIX systems, the Digidesign *universal slave driver* (USD) provides this function. For HD systems, the Sync I/O is the correct device. The USD or Sync I/O can be locked to an external reference signal or can act as the master timing device. The exact protocol used is determined in the software setup. The Digidesign manuals provide extensive detail on this subject.

There are also synchronization devices and combination MIDI/synchronization devices available from a number of third-party vendors, including MOTU and Emagic.

Control Surfaces   The Pro Tools interface, particularly the Mix window, is designed to intuitively emulate physical devices. Although this emulation generally works very well, some operations are hard to perform with a mouse. Perhaps the greatest weakness of the system in this regard is the inability to move multiple faders independently with a mouse. Fading one track up while fading another track down when you only have one mouse to control the faders is basically impossible. It becomes a two-pass operation.

That's where control surfaces are useful. A control surface is a device that provides tactile control over the software. Most control surfaces have multiple physical faders that can be linked to the virtual faders in the Pro Tools Mix window and be used to control the software faders simultaneously. It basically makes the experience of mixing in Pro Tools more like mixing on a console in a traditional audio recording and mixing environment. Most control surfaces provide a range of controls, such as transport controls and the ability to adjust the parameters of filters.

The only Pro Tools system that comes with a control surface is the Digi 002. Digidesign has two other control surfaces that work with MIX and HD systems, which are purchased separately. They are the Control 24 and the Pro Control. These devices are both very sophisticated pieces of equipment that provide a wide range of functionality. Control 24 is a 24-track, self-contained device. Pro Control is a modular system that starts with a unit containing eight faders, to which additional units of eight faders each can be added. There is also a special Edit Pack module for Pro Control that provides additional controls for surround mixing and other functions. There are a lot of third-party control surfaces available as well. Some of them are specific to Pro Tools and some are capable of controlling not only Pro Tools, but other software as well. Control surfaces connect to the computer via MIDI, Ethernet, USB, or FireWire connections, depending on the model.

Mixers   There are several reasons why you might want to have an external mixer for a Pro Tools system. If you have a lot of input devices, a mixer can be used as a router.

The microphones, instruments, CD or tape players, or other devices can be routed into a mixer. The output of the mixer can then be routed to Pro Tools. This method is a common way of setting up a project-studio environment with Mbox, Digi 001, or MIX systems. The use of a mixer in these kinds of environments can reduce the need for a lot of Pro Tools I/O interfaces. A mixer can also adjust and control the signal levels from various sources prior to recording them to Pro Tools.

Some digital mixers have extensive built-in digital effects and are MIDI capable. These additional effects can be added to the incoming signal at the mixer stage. Similarly, some external signal processors and effects modules can also be controlled via MIDI. Applying these effects externally preserves Pro Tools' processing power and adds to the overall flexibility of the system. However, the addition of a MIDI-controlled external module also increases the system's complexity. Weighing the benefits of the additional power versus the cost of the additional complexity can be important in system design.

Microphone Preamps   Microphone preamps are necessary because most microphones produce a relatively low signal amplitude. The microphone preamp is used to boost the signal up to the line level. This boost can be up to 60 dB. Sixty dB is a huge amount of amplification, much more than the signal will experience throughout the rest of the subsequent recording and mixing. Many engineers agree that good mic preamps, while sometimes very expensive, are one of the best investments that can be made in a recording system.

A mic preamp should have a very high signal-to-noise ratio, on the order of 90 dB or more, in order to be able to provide the needed boost to the mic signal without adding undue noise. It should also have a very flat frequency response and be sufficiently shielded to prevent significant RF interference.

Many mixers have microphone preamplifiers built into their channels. The quality varies from very bad to very good. These built-in preamps are often the most cost effective microphone preamps and can provide the benefit of a good-quality preamp without adding another piece of gear.

However, there are also many high-quality single- and multiple-channel standalone microphone preamps available. The designs and sounds of preamps vary considerably. Some are solid state and some use tubes for amplification. There are also hybrid designs that incorporate both tube and solid-state components. There is no decisive opinion on which is better. Many engineers feel that the tube models produce a warmer sound. However, there are also those who favor the possibly cleaner sound of transistor-based units. At the higher end of the spectrum, this difference becomes a matter of personal judgment and taste.

The bottom line is that a good preamp will make the incoming microphone signal sound better in a number of ways, depending on the model. The better the incoming signal, the less you will have to adjust it in the mix. Good microphone preamps can cost from a few hundred to several thousand dollars.

Digidesign makes a microphone preamp called Pre. Pre is a rack-mount unit that was introduced along with the HD systems hardware. It has eight separate preamplifier channels with both XLR and ¼-inch phone jacks that can accept mic, line, or instrument levels. It has MIDI-in, MIDI-out, and MIDI-thru and can be controlled from Pro Tools software.

INDEX

Note: Boldface numbers indicate illustrations.

N

O

P

ABOUT THE AUTHOR

David Leathers is a musician and filmmaker. After working as a studio and touring musician in New York during the '70s and early '80s, he relocated to Los Angeles and was trained in audio recording engineering and video production. For the last 20 years he has been wearing a variety of hats in video and film production, postproduction, and music. He's also worked extensively as a writer and consultant in the field of audio and video technology and computer-based systems. He's reviewed hundreds of products for video and audio magazines, including *Videography, EQ, Video Systems, Broadcast Engineering, Digital Cinema,* and many others. He started working with Pro Tools systems about nine years ago in the context of sweetening audio for video projects and now uses the system for both music recording and audio post in his Los Angeles studio.